Smarter Than the Storm

Praise for *Smarter Than the Storm*

'The rapid ascent of artificial intelligence presents the global energy sector with two major developments: a significant new source of electricity demand on the one hand, and a powerful new tool that could optimize energy systems and find innovative solutions to complex climate challenges on the other. This book provides a timely exploration of these complex dynamics, offering a compelling and clear-sighted view of the interplay between AI, energy and sustainability.'

— **DR FATIH BIROL,** EXECUTIVE DIRECTOR, INTERNATIONAL ENERGY AGENCY (IEA)

'This is a book of crucial importance from authors who have been at the heart of creating action at the national and international level, both in responding to the climate crisis and in building a new story of growth and development, which is sustainable, resilient and inclusive. Much more attractive than the dirty, destructive models of the past. Innovation for clean and efficient activities and infrastructure will be critical. Investment will be at centre stage. So too, AI. They also identify clearly and strongly international obligations, particularly of the rich countries who followed a polluting model of development. This is a very valuable contribution to international understanding and debate.'

— **LORD NICHOLAS STERN,** FBA, FRS, CH, KT., IG PATEL PROFESSOR OF ECONOMICS AND GOVERNMENT, CHAIR OF THE GLOBAL SCHOOL OF SUSTAINABILITY, CHAIR OF THE GRANTHAM RESEARCH INSTITUTE ON CLIMATE CHANGE AND THE ENVIRONMENT, THE LONDON SCHOOL OF ECONOMICS AND POLITICAL SCIENCE

Smarter Than the Storm explores how climate change is shaping the future of technology and how artificial intelligence can help address the climate crisis without exacerbating environmental pressures. As the digital and climate transitions become increasingly intertwined, the book offers a timely and valuable reflection on how aligning these transformations can accelerate sustainable development in the Global South, advance the Paris Agreement's long-term goals, and support the fight against poverty, hunger and inequalities.

— **AMBASSADOR ANDRÉ CORRÊA DO LAGO**, PRESIDENT of COP 30

'*Smarter Than the Storm* shows how the AI revolution and the climate crisis can be tackled together rather than separately. It offers a rare combination of clear analysis, institutional realism and evidence-based policy proposals for emerging and developing countries. A valuable roadmap for governments, investors and multilateral lenders.'

— **PROFESSOR ANDRÉS VELASCO**, DEAN, SCHOOL OF PUBLIC POLICY, THE LONDON SCHOOL OF ECONOMICS AND POLITICAL SCIENCE, AND FORMER FINANCE MINISTER OF CHILE

'*Smarter Than the Storm* is a compelling exploration of the feedback loop between climate change and AI. Amitabh Kant and Siddharth Sinha trace how climate shocks can trigger resource scarcity and conflict—and how AI, despite its own environmental footprint, can become a powerful force for mitigation and resilience. With fresh ideas from sovereign AI to community-positive data centres, this book reframes challenges as opportunities and offers a bold, practical roadmap for sustainable and scalable climate action.'

— **EMMA SKY, OBE**, DIRECTOR, INTERNATIONAL LEADERSHIP CENTER, YALE UNIVERSITY

'*Smarter Than the Storm* is a confident, forward-looking exploration of how humanity can meet climate risk with intelligence rather than fear. Clear-eyed about the scale of disruption, the book reframes climate as a threat multiplier and AI as a powerful force multiplier, capable of strengthening resilience, accelerating decarbonization and advancing equity. With a strong systems lens, it connects technology, finance, governance and behaviour into practical pathways for action, particularly for emerging economies. Like Odysseus steering by insight rather than impulse, the authors argue that wisdom, collaboration and openness matter more than brute force. Accessible yet rigorous, this is a compelling call to act boldly, collectively and now.'

— **MALCOLM MCCULLOCH,** PROFESSOR AND GROUP LEADER, ENERGY AND POWER GROUP, UNIVERSITY OF OXFORD

'*Smarter Than the Storm* offers a compelling look at how climate change and AI will shape our future. It makes complex topics accessible for everyone, from seasoned practitioners to curious readers. By telling this story through characters that reflect lived realities, the book invites us to think boldly about climate solutions that are both effective and people-positive.'

— **VRUSHALI GAUD,** GLOBAL DIRECTOR, CLIMATE OPERATIONS, GOOGLE

Smarter Than the Storm

Championing the
AI-Climate Nexus
for a Truly Sustainable Future

Amitabh Kant | Siddharth Sinha

HarperCollins *Publishers* India

First published in India by HarperCollins *Publishers* 2026
HarperCollins *Publishers* India, Cyber City, Building 10-A,
Gurugram, Haryana-122002, India
www.harpercollins.co.in

2 4 6 8 10 9 7 5 3 1

Copyright © Amitabh Kant and Siddharth Sinha 2026

Graphic Art Credit: Manjari Medhavee, Twig Designs

P-ISBN: 978-93-6989-351-5
E-ISBN: 978-93-6989-813-8

Amitabh Kant and Siddharth Sinha assert the moral right to be identified as the authors of this work.

All rights reserved. No part of this publication may be reproduced, stored in a retrieval system, or transmitted, in any form or by any means, electronic, mechanical, photocopying, recording or otherwise, without the prior permission of the publishers.

Without limiting the exclusive rights of any author, contributor or the publisher of this publication, any unauthorized use of this publication to train generative artificial intelligence (AI) technologies is expressly prohibited. HarperCollins also exercise their rights under Article 4(3) of the Digital Single Market Directive 2019/790 and expressly reserve this publication from the text and data-mining exception.

Typeset in 12/16 Adobe Caslon Pro
by HarperCollins *Publishers* India Pvt. Ltd

Printed and bound at
Manipal Technologies Limited, Manipal

This book is produced from independently certified FSC® paper to ensure responsible forest management.

HarperCollins *Publishers*, Macken House, 39/40 Mayor Street Upper, Dublin 1, D01 C9W8, Ireland

CONTENTS

Prologue ix

Part 1—Climate

1. An Uncertain World
   Crisis, Chaos and Climate 3
2. A Warming World
   A Primer on Climate Change 14
3. Tech on Thin Ice
   When Technology Gets Caught in the Climate Crossfire 27
4. The Geopolitics of Climate Change
   Is Climate Threatening the World Order? 36
5. Rethinking Climate Equity
   Humanity's Shared but Differentiated Burden 47
6. Challenges as Opportunities
   Leapfrogging to a Sustainable Future 54
7. On the Cutting Edge
   Climate Technologies Enabling Transition and Shaping a Greener Tomorrow 59

Part 2—AI × Climate

8. The AI Revolution
   Powering Humanity's Greatest Leap   77
9. Predicting the Unpredictable
   AI for Climate Resilience and Adaptation   91
10. Decarbonizing the Planet
    AI for Climate Mitigation   110
11. Intelligence at Scale
    AI for Systems Transformation and Climate Governance   130
12. From Dilemma to Design
    Building Scalable, Sovereign and Sustainable AI for Climate   143

Part 3—The Blueprint

13. Data Centres for the Future
    Building Green Data Hubs, Empowering Communities   155
14. DPI and Data Commons for Climate AI
    India's Digital Inclusion Playbook,
    Rebooted for Democratizing Climate Data   176
15. Decolonizing AI
    Championing Sustainable Development Through Open
    and Sovereign AI Solutions   195
16. A Systems Lens
    A Pragmatic Policy Framework for Approaching
    the AI–Climate (or Any Other) Nexus   210
17. Funding the Future
    Financing AI-Driven Climate Innovation and Action   227
18. Closing the Loop
    Making Sense, Moving Forward, Together   244

Epilogue   261
Notes   265

PROLOGUE

We stand at a pivotal moment, defined by immense challenges but also extraordinary opportunities.

You do not need to be an expert to read this book. And this is not your typical climate or AI book in a world overflowing with them. Whether you are a student, a professional, a researcher, a policymaker, or simply someone curious about the world around you, and how it is changing, this book offers a clear, accessible exploration of the climate crisis, the rise of artificial intelligence, and the vital yet still underexplored space where they intersect. With a multidisciplinary and multidimensional lens, this book shows how their convergence can unlock a more resilient, sustainable and flourishing planet.

At the heart of this narrative lies climate change, the greatest threat multiplier of our time. It deepens vulnerabilities, fuels conflicts, disrupts resources essential for life and progress, and challenges the systems that support billions. It creates *vicious cycles* that are increasingly hard to break. And its effects spare no domain. From triggering extreme weather events that heighten geopolitical tensions, to disrupting critical mineral supply chains, to affecting everything from energy grids to flight turbulence, climate change has not even spared technological advancement, which we desperately need to transition to a net-zero world. Its impacts cut across the Global North and the Global South, yet those who bear the least responsibility are often the most exposed. Today, we stand at a critical and time-sensitive juncture. Unlike the developed world—which industrialized through coal and high-emission pathways—emerging economies now have access to clean, cutting-edge technologies that never existed before. These range from solar energy to green molecules. The question is whether these nations can leverage these innovations to industrialize sustainably and prosper, while avoiding the destructive path of the past.

Yet, amidst the complex and pressing challenges brought about by climate change, an extraordinary opportunity emerges in the form of artificial intelligence (AI). Yes, AI is one of the most overused buzzwords of our time, but in this book we explore a very different side of it. When aligned with climate technologies, AI becomes a genuine force multiplier that can accelerate climate action, strengthen adaptation and resilience, and transform how we understand and manage the challenges ahead. But AI is not without its own footprint. The systems we increasingly rely on consume enormous amounts of energy and water, and demand vast volumes of high-quality data that are still fragmented and inaccessible. Many advanced AI models remain closed, placing limits on innovation and national capability. These tensions mirror the very complexity of the climate challenge itself.

But instead of seeing these issues as roadblocks, this book frames them as powerful opportunities. They point us towards transformative solutions: green data centres that cut emissions while creating value for local communities; digital public infrastructure that unlocks climate data for innovation; and open, sovereign AI models that nations and citizens can shape, govern and trust. Realizing this vision requires strong ambitions matched by strong action. It demands policy frameworks with teeth, financing mechanisms that scale what works and discard what does not, and global cooperation rooted not in rhetoric but in results. These are the levers needed to unlock the full potential of the AI-Climate nexus and catalyse lasting systemic change.

As we explore these themes, we invite you to follow the journey through the lives of characters whose stories are intertwined with the narrative of this book. They offer a human lens on the complex forces shaping our world and the possibilities that lie ahead. Throughout, we simplify the interconnected threads of climate change, technology, geopolitics and AI to reveal how these forces are deeply linked. The book shows that challenges and opportunities are not opposites but part of the same system, and that taking a 360-degree view is essential if we are to design solutions that actually work.

The time to act is now. In our own lives, in our work and in the choices we make, we often see problems and possibilities in silos. But the climate and AI systems shaping our century do not operate in silos. And if we want to break the loops that hold the world back, the moment to rethink and reimagine our path is right here in front of us. Ultimately, this book is not just about global systems or distant futures. It is about you, our world and the possibilities that open up when we choose to shape the century rather than be shaped by it.

PART 1
CLIMATE

1

AN UNCERTAIN WORLD

Crisis, Chaos and Climate

A WORLD ON THE EDGE

Somewhere above a conflict zone, a fighter jet cuts through the sky. It is barely visible, but its consequences are deeply felt. The sound of uncertainty across the world today, if it had one, would drown even the sonic boom that the jet produced. Asha, a fifteen-year-old girl who has been waiting for over six hours in a queue to collect relief supplies along with her mother in a village, looks up at the contrails left by the jet in the sky. A prolonged drought has caused a shortage of food and water, and a fight has broken out between two armed groups over the limited supplies of food and water that have remained. Many innocent villagers have lost their homes. Asha's father, accompanied by her younger sister, has moved to the city to find an informal job to help his family get back on its feet. Asha's name carries two meanings: in Swahili, Asha evokes life, and in Hindi, it embodies hope. In this moment, both life and hope are all her family has left to hold on to.

Several thousand kilometres away, in the Western world, Sophie, a fifteen-year-old girl, is playing in her backyard. She goes to school, lives in a comfortable home and her parents run a local bakery. They have never experienced a shortage of anything: food, water or other basic necessities. It is an unusually hot Sunday afternoon, and her parents casually remark on how warm it is getting these days. It feels like a regular summer day, perhaps just a little too hot for the season. No one gives it much thought. But two days later, Sophie and her family find themselves in temporary tents inside a sports stadium. Their home, along with the rest of the neighbourhood, has been reduced to ashes after a wildfire swept through without warning. The unusually warm weather had acted as an accelerant, fuelling the blaze that changed everything in moments. Her country had also recently withdrawn from a key climate pact, shockingly dismissing climate change as a hoax, even as heatwaves and wildfires scorched its own landscape.

On another side of the world, sudden floods affect the production of semiconductors, disrupting—albeit for a short time—the supply

chains that power everything from phones to cars, and power plants to electric vehicles. It sends shockwaves through global markets, exposing just how vulnerable many critical industries remain to single points of failure and that climate change is increasingly exposing these points of failure. This sparks urgent conversations around the world about the need for greater resilience and self-sufficiency.

And just as the dust begins to settle from one crisis, another is already taking shape quietly, often even before the last one has fully ended. It could be economic, environmental, geopolitical, or as is now increasingly common, a mix of all three unfolding at once. Does this seem like fiction? Actually, it isn't. Depending on where you are in the world, it either already is or is rapidly becoming the lived reality of our times. From conflicts and trade wars to policy volatility and an accelerating climate crisis, the world today faces a cascading series of crises.

You could be one of the many who have already lived through this. Maybe you were forced to evacuate your home during a sudden flood. Or perhaps you lost your job as your company struggled to cope with rising interest rates, economic uncertainty, global instability, or even the growing adoption of AI that is automating a number of roles. If you are in a certain part of the world, a conflict may have driven up oil prices, making electricity and heating painfully expensive. You find yourself watching your savings shrink just to keep your home warm.

Even if you consider yourself relatively fortunate, chances are you have still felt the undercurrents of this uncertainty. Perhaps your morning commute takes longer than it used to. Sudden downpours are becoming more frequent, breaking away from the patterns you had grown used to over the years. Sometimes it is the rain, at other times it is the heat causing electrical breakdowns that delay public transport. You have started noticing that your usual shopping, even for groceries, feels heavier on the wallet, not just because of rising prices, but also because fruits and vegetables arrive late or in smaller quantities. Your electricity bill has crept up, and you find yourself using the fan or the heater more often than before because of how hot the summers have become. If you are a parent, your child's school now sends weather alerts more frequently, and unexpected days off due to storms or heatwaves have become part of routine. Today, perhaps even to plan a short holiday, you find yourself checking weather updates, transport schedules and fuel prices, repeatedly adjusting your plans. If this is the amount of uncertainty involved in something as simple as a weekend getaway, imagine what it takes to plan something long-term like buying a new house.

For many of us, history played out in the pages of textbooks. Grainy black-and-white photographs captured wars of the past, while pandemics like Black Death—the plague that wiped out millions—were portrayed through images of hooded figures carrying burning torches through streets lined with lifeless bodies. If you are part of the younger generation reading this book, you may never have imagined that you would live through a war or a pandemic. Yet, in just the past few years, the world has witnessed not one, but many wars. We have lived through a global pandemic that led to grounded flights, locked-down cities and millions of people losing their lives. It has exposed the fragility of systems we once assumed were unshakable. Climate change used to be something we talked about in science class, wrote about in essays and debated in school competitions. It felt distant, reaching our classrooms through images of cracked ice shelves in the Antarctic or a lone polar bear drifting on a shrinking patch of ice. But now it's at our doorstep. Look around.

Today, we are at a point where growth projections have been cut for almost 70 per cent of countries across the world. According to the World Bank's latest Global Economic Prospects report,[1] global growth is projected to slow down to its weakest pace since 2008, excluding periods of full global recessions, largely due to rising trade tensions and ongoing policy uncertainty. This is troubling news for a world still recovering from the aftermath of COVID-19, grappling with ongoing conflicts, and facing the growing and increasingly visible threat of climate change.

EDGES THAT FRAY FIRST

While rebuilding a house for Sophie's family will be tough, it would be near impossible for Asha's family. Asha and her family represent the millions, if not billions, of people living on the edges of society—poor, marginalized, vulnerable and displaced. These are people who have been left behind by faltering economies, ignored by society, displaced by conflict, or left homeless because of floods, fires, wars and failing states. If you are poor and vulnerable, the impacts will hit you harder. With income or social support, you can still absorb some shocks such as repairing a damaged roof, moving to a safer place or paying a medical bill. But without it, even small disruptions can spiral into hardships, pushing people into poverty traps that are self-perpetuating and nearly impossible to escape. When you live on the margins, there is no cushion to fall back on, no safety net to catch you and no easy way to recover. Slowing growth will further hinder developing economies in their efforts to create jobs and reduce extreme poverty while widening the income gap with advanced economies, which are themselves not immune to the effects of this slowed-down growth.

Globally, anywhere between 1.3 to 1.5 billion adults today remain unbanked, and many more lack access to digital financial services. This makes life significantly harder during times of crisis and economic disruption. Would it be possible to get by without a debit or credit card, or other forms of digital payment, today? No mobile apps, no online transfers, no way to tap and pay.

Now imagine not even having a bank account to begin with. Over 700 million people across the world today live in extreme poverty, which is almost 8–9 per cent of the world's population. If a recent study by the United Nations Development Programme and the Oxford Poverty and Human Development Initiative is anything to go by, over 1 billion people (or 18 per cent of the world's population) live in multidimensional poverty.[2] Multidimensional poverty is perhaps a more realistic way of measuring poverty, especially since it takes into account disadvantages beyond lower incomes such as lack of access to health, education and so on.

The Sustainable Development Goals (SDGs), adopted by all United Nations Member States in 2015, are a set of 17 interconnected goals designed to serve as a blueprint for achieving a better and more sustainable future for all by 2030. They cover a wide range of global challenges, from poverty and hunger to education, health, gender equality, climate action and peace. The UN's Sustainable Development Goals (SDG) Report 2024 tells us that only 17 per cent of SDG targets are currently on track, while nearly half show only minimal or moderate progress.[3] Worryingly, progress on over one third of the targets has stalled or even regressed. The report highlights how the lasting impacts of the COVID-19 pandemic, intensifying geopolitical conflicts, rising climate chaos, and global economic inequalities are hitting SDG progress hard. Structural deficiencies in the global economic and financial systems are leaving developing countries with just a fraction of the support they need to tackle mounting challenges. Inequalities continue to widen, biodiversity loss is accelerating and ongoing conflicts have caused displacement on a large scale. These statistics matter because while a delayed train, a rising electricity bill or a flooded street may be an inconvenience for some, they represent existential challenges for those already struggling to survive.

EDGE THAT CARRIES THE WEIGHT: GLOBAL SOUTH

Developing countries, in particular, feel the weight of these converging and cascading crises more acutely than others. They have low per-capita incomes, and while some of them are growing very fast economically, many face poor medium-term economic outlooks and suffer from high debt burdens and inadequate access to finance. Social safety nets in many of these countries are limited, and even basic services such as healthcare, clean drinking water, reliable electricity and safe housing remain out of reach for large segments of the population. Urban centres are often overburdened and many people live in high-risk zones that are especially vulnerable to climate-related disasters—and not by choice—they simply cannot afford to live elsewhere. These countries are not only contending with the lingering effects of the pandemic, disruptions caused by conflicts, a paralysis of decision-making at international organizations, but also face disproportionate exposure to the impacts of climate change. A term that is increasingly used to describe this group of countries is the 'Global South'. While the definition is not strictly geographic, it generally refers to developing and emerging economies across Asia, Africa, Latin America, and parts of Oceania. If one were to look at the map of the world, this distinction loosely aligns with geography. Most countries in the Southern Hemisphere fall under this category, though there are notable exceptions. Australia and New Zealand, for instance, are geographically in the south but do not face the same challenges as their regional neighbours. Hence, they do not fall into the 'Global South' grouping. Similarly, some countries located in the Northern Hemisphere, such as India or Egypt, are still considered part of the Global South because they share similar developmental characteristics and constraints.

What unites many of these countries is a complex combination of shared vulnerabilities and common aspirations. Populations in the Global South are often more exposed to climate change, food insecurity and economic volatility. Their health and education systems tend to be more fragile, and their economies more dependent on sectors vulnerable to global disruptions. But these countries also share important strengths. Many of them are fast-growing economies with rich cultural capital, natural resources, and most importantly, young and aspirational populations who are eager to shape a better future. While the Global South may be disproportionately affected by today's global challenges and uncertainties, can it turn these challenges into transformative opportunities? As you go through this book, you might start to see that it is not just possible, but many countries have already taken the lead. But a lot more needs to be done, and not just by the Global South. The Global North has a big role to play too.

LIVING ON THE EDGE

Asha's father moved to the city with her younger sister to help the family get back on their feet. Asha and her mother stayed back in their village to support their ailing grandparents. Her father is now a daily wage labourer living in a crowded, low-income neighbourhood on the outskirts of a capital city in a country in the Global South. The city is reeling under stress from massive migration from rural areas as unprecedented drought conditions prevailing for the past three years have decimated fields of crops. Rent takes up most of Asha's father's income, he has no health insurance, barely any savings, and his job depends on a long commute traversing many modes of transport—a bus, then a train and then a tuk-tuk. In a country otherwise facing drought and where the capital city has never known incessant rains, a continuous downpour lasting for days has flooded the city. This is not completely unusual in a world where weather patterns have become extremely unpredictable due to climate change. The poorly maintained drains are not designed to withstand such a massive amount of stormwater, and have overflowed. Her father's house has been completely submerged, and everything inside is ruined. The appliances, the mattresses and even the small savings tucked beneath the sofas are lost. Built in a slum on weak foundations, their house collapsed on the third day of the floods as the ground under the foundations sank. As the floodwaters stagnated, it presented a breeding ground for mosquitoes, which led to dengue. Asha's sister fell sick and needed to be rushed to a hospital, but the government hospitals had run out of space. Private hospitals had a few beds, and Asha's father had to borrow money from an informal lender (who charged exorbitant interest rates) to get her hospitalized. He could not get to work for a week as public transport services were no longer running. When he did, the small shoe factory that he worked at was under water, their machinery ruined. He now had no job. The local clinic was overcrowded and he struggled to buy medicines for his daughter. He developed diarrhoea from drinking unsafe water. Vegetables were more expensive because of the drought and the ongoing conflict in the countryside. Besides, a global trade war had also meant additional tariffs on their country, making some essential items even more expensive. He kept borrowing money from the ruthless moneylender just to keep the household running. The family was already living in poverty. They earned barely enough to survive, managing only the most basic shelter and just enough food, which was far from nutritious.

Their home was fragile, located in a crowded area with poor drainage and sanitation, in a city with an overwhelmed health system. Every aspect of their lives was already under threat. They faced unstable income, poor health, inadequate sanitation, fragile housing and limited access to education. Then came *something* that did not just add to their hardships but intensified them. It triggered heavy rains that spiralled into a public health crisis, a job crisis and a housing crisis. It took each of these existing vulnerabilities and made them worse. This was not just a threat. It was a threat multiplier. It affects Asha, it affects Sophie and it affects the world.

Already, conflicts coupled with the pandemic have drained resources. Trillions that could have gone into development, education, health, infrastructure or innovation, have instead gone into recovery and rebuilding. Today, policy uncertainty, trade wars and economic instability are only making the situation worse. These are all deeply interconnected threats that constitute the cascading crises we face today. Their combined impact is already severe. When *something*, an additional threat, enters the equation, it intensifies every existing issue, turning an already difficult situation into one that is devastating.

But what is this *something*? Look around you.

2

A WARMING WORLD

A Primer on Climate Change

That *something* is climate change.

SKY

On 21 May 2024, Flight SQ321 from London Heathrow to Singapore encountered severe turbulence over Myanmar, leading to many injuries and one fatality. Photos and videos of the incident quickly went viral on social media, but it was only one among many reported cases of increasing turbulence on flights around the world. The airplane fell victim to clear air turbulence (CAT), a phenomenon which is completely unpredictable. If you are particularly nervous about turbulence on airplanes and have started noticing more such events, know that climate change could be a culprit. Research shows that rising atmospheric temperatures have led to an almost 55 per cent increase in clear air turbulence on airplanes between 1979 and 2020.[1]

When we said in the last chapter that climate change is no longer something you just read about in textbooks, win debate competitions over, or associate with distant melting glaciers, this is exactly what we meant. It is beginning to surface in our daily lives, in places and moments we never imagined. If you thought roads in developing countries were bumpy, air travel across the world could now be bumpier—so buckle up.

Unlike a war that may be limited in some ways by geography or time, climate change is omnipresent. You can stop a war in minutes after a ceasefire agreement or the signing of a peace treaty. You cannot do that with the climate. The effects of climate change do not spare any part of the world, whether you live in Asha's part of the world or Sophie's.

From the fires in Canada to the floods in Bangladesh, from the rising sea levels in the Pacific Islands to droughts in southern Africa, climate change is everywhere. It affects everyone. It compounds every crisis. Its impact cuts across every sector.

EARTH

You do not need to be a researcher or a scientist to find evidence of climate change and the effects that it has been having on people and the planet. If you noticed that you were sweating and panting quite a bit during the last few summers, that is because they were hotter and longer than usual. It is quite likely that what you are witnessing are the impacts of climate change in the form of extreme heat. Many houses in Europe never had fans and air conditioners (ACs) to begin with, and yet, today, sales of air conditioners and fans in European cities is booming. New houses are being built with air conditioning installed. A study published in *Energy Research & Social Science* indicates that the number of UK homes experiencing overheating rose from 18 per cent in 2011 to 80 per cent in

2022.² This surge has led to a sevenfold increase in the use of air conditioning. Up until a few years ago, Sophie's family never needed to use ACs. Their house, which burned down in the wildfire, had four of them installed two years ago at the peak of an unprecedented heatwave. In Europe, temperatures are soaring at twice the global average and studies estimate that over 47,000 people died in the continent in 2023 due to the heat.³ For the summer of 2024, the estimate is about 62,775 heat-related deaths, an increase of over 23.6 per cent compared to 2023.⁴ The study goes on to show that the number of fatalities would have been almost 80 per cent higher in the absence of the adaptation measures that have been undertaken today.⁵ Air conditioners do save lives, but they also fuel global warming by increasing energy consumption. If this energy is not coming from renewables, then it contributes directly to global warming. Air conditioners also expel hot air which causes outdoor temperatures to further soar. If you are an Indian navigating the crammed bylanes of congested Delhi neighbourhoods, or a foreign vlogger capturing the chaos of its narrow alleys (this seems to have become quite a trend of late) during peak summers, thousands of outdoor AC units, hanging precariously from window sills, will only add to your sweat. Excess heat generated by a city's worth of air conditioners can increase the outside temperature by 1 to 1.5 degrees Celsius. This is a less-than-ideal situation for cities already buckling under the effects of heatwaves. A recent study found that heat-related deaths in India increased by about 55 per cent when comparing the period 2000–2004 to 2017–2021.⁶ Mexico, Thailand, and West and Central Africa all experienced widespread heat-related events, with the latter reporting thousands of deaths due to prolonged high temperatures.

The world is becoming too hot to live in, and unless our response is rooted in alternative sustainable solutions, we will only make the problem worse. Global warming causes temperatures to rise, we then try to cool our homes and offices by installing ACs that consume vast amounts of energy, and in the process, we cause more global and local warming. The human race has trapped itself in a *vicious cycle*: You will meet this term again in this book, more than once, and in time, you'll see how they have been quietly reshaping life as we know it.

The impacts of climate change are visible across all sectors. About 11 per cent of Earth's land surface is used to grow crops. However, erratic weather patterns, prolonged droughts and extreme heat events are drastically affecting crop yields, leading to food insecurity in many regions. Farmers are struggling to adapt to changing growing seasons, reduced water availability, and increased vulnerability to pests and diseases, further exacerbating the already fragile food systems. As per a NASA report,[7] maize yields could decline by as much as 24 per cent, whereas wheat could see a 17 per cent decline by 2030. At COP28, a report[8] released by the Food and Agriculture Organization (FAO) stressed that agrifood systems are intrinsically linked to climate change and are particularly vulnerable to its impacts. Each year hundreds of billions of dollars' worth of crops and livestock production is lost due to disaster events.

Infrastructure, too, is increasingly under siege from climate change, with extreme weather events such as floods, wildfires and storms wreaking havoc across the world causing loss of lives and inflicting widespread damage on homes, roads, bridges, water systems and digital networks. For countries in the Global South, this destruction compounds existing challenges: balancing investments in basic infrastructure and poverty alleviation with the urgent need for climate adaptation. Consider a scenario where a developing nation, already striving to meet the basic needs of its citizens, initiates investments in renewable energy infrastructure. Soon after, a climate-induced extreme weather event devastates the region, destroying not just their renewable energy infrastructure but also homes and hospitals. A lot of funds that could have been used to combat climate change and increase investments in renewable energy would now be urgently used to aid rebuilding of cities, crowding away funds from climate projects, thus slowing efforts aimed at combating climate change. This is the same *vicious cycle* mentioned earlier at work.

The Coalition for Disaster Resilient Infrastructure (CDRI) reports[9] that over USD 2 trillion worth of infrastructure assets in South Asia are exposed to climate risks, with 30 per cent attributed to the transport sector alone. Proactive measures in climate-resilient infrastructure are no longer a matter of choice—they are a necessity. The Global North is also buckling under these effects. The European Environment Agency warns that without adaptation, economic losses from climate-related events like heat waves and floods could reach up to €1 trillion annually by the end of the century.[10]

This interplay between climate change and infrastructure exposes yet another *vicious cycle*: climate events damage critical infrastructure, hindering development and adaptation efforts, which in turn exacerbates vulnerability to future climate impacts.

AIR

Air pollution is a major global health crisis, and if the estimates of the World Health Organization are to be believed, it causes an estimated 7 million premature deaths each year.[11] As per the World Air Quality Report of 2024,[12] only about 17 per cent of cities globally meet the WHO guideline for fine particulate matter or PM2.5. PM2.5 refers to fine particulate matter smaller than 2.5 micrometres that are tiny enough to enter deep into the lungs and even the bloodstream. But air pollution goes far beyond PM2.5. PM10 includes slightly larger particles that still penetrate the respiratory system. PM1, even smaller than PM2.5, can travel deeper into the body and is linked to severe cardiovascular and neurological risks. VOCs (volatile organic compounds) such as formaldehyde are gases released from vehicles, industries and even household products, and they can irritate the lungs, trigger asthma and contribute to long-term health damage. Together, these pollutants form a dangerous mix that harms health, reduces life expectancy and burdens communities everywhere.

It is interesting to observe the many ways in which air quality is linked to climate change. Poor air quality is driven by emissions along with other contributors such as dust. These emissions, which come from cars, buses, motorbikes and factories, not only degrade air quality, but also contribute to climate change. Climate change, in turn, alters weather patterns in several ways, including changes in rainfall cycles, rising temperatures, and the occurrence of stagnant air conditions. Reduced rainfall leads to higher pollution levels, as rain helps clean the air by picking up pollutants hovering in the atmosphere and washing them out. Stagnant air traps pollutants within cities, while colder winters can cause polluted air to stay closer to the ground, worsening what people breathe. Rising temperatures also increase ozone formation, a harmful pollutant that contributes to smog. Warmer conditions also fuel more frequent and intense wildfires, releasing large amounts of particulate matter and toxic pollutants into the air, degrading air quality. Children are especially vulnerable to the impacts of polluted air, which can impair lung development and increase the risk of lifelong respiratory illnesses. Almost 700,000 children under five years of age lose their lives to air pollution.[13] Beyond health, the economic toll is staggering: the World Bank estimates that air pollution costs the global economy over $8.1 trillion annually, or 6.1 per cent of global GDP, due to lost productivity, healthcare costs and premature deaths.[14]

A question that we must ask ourselves is this: can we continue to build a world in which basic human rights are a privilege enjoyed by a few? How can we ensure that emerging economies are able to sustain high-growth trajectories while safeguarding the health and well-being, and in turn, the latent potential, of their populations?

FIRE

The rise in wildfires is another alarming consequence of climate change, with both frequency and intensity climbing worldwide. In Australia, some 10 million hectares were lost in the 2019–20 Black Summer fires. Three years later in 2023, the blazes in the north of the country scorched at least 84 million hectares, which is roughly the size of Germany, marking what some experts call the largest fire season in decades. In the United States, especially in California, wildfire seasons continue to grow in threat and scale, with over 1 million acres burned in 2024. In India, states like Uttarakhand have recorded more than 21,000 fire events from late 2023 to mid-2024, impacting thousands of square-kilometres of forests and placing rural mountain communities in peril. These fires are fuelled by rising temperatures and prolonged droughts, which are becoming more common due to climate change. This is particularly concerning as wildfires not only contribute directly to emissions by releasing large amounts of carbon dioxide (CO_2) into the atmosphere, but also destroy the natural ecosystems that help absorb this very CO_2.

Forests play a crucial role in sequestering carbon, and when they burn, this vital function is lost, creating a pattern where increasing emissions from wildfires further exacerbate climate change, leading to more frequent and intense fires in the future—birthing another devastating *vicious cycle*.

WATER

The effects on water are equally concerning, and ironically, both scarcity of water and a catastrophic excess of it (floods) are on the rise. Higher atmospheric temperatures and increased water vapours due to climate change intensify extreme weather events such as torrential rain and flooding. At the same time, changing rainfall patterns, prolonged dry spells, and faster evaporation caused by rising temperatures can lead to water scarcity.

High water stress is likely to affect about 250 million people in Africa and is expected to displace up to 700 million people by 2030, as per the State of Climate Report 2021 for Africa by the World Meteorological Organization (WMO).[15] India's average annual per capita water availability is projected to decline from an already-stressed 1,486 cubic metres in 2021 to 1,367 cubic metres by 2031, according to India's Ministry of Water Resources. This places the country well below the water stress threshold of 1,700 cubic metres, edging closer to 1,000 cubic metres, which is the benchmark for water scarcity.[16]

Our relationship with water is a factor of timing and scale; as a civilization, we have sought water to set up communities, to sustain agriculture, and to power industries. Throughout history, we have tried to control it, including engineering dams, channels and irrigation systems to predict its flow and harness its power. Water is both a friend and a foe; it nurtures life and fuels growth, yet when it overwhelms us in floods or vanishes into droughts, it becomes a destructive force that threatens our existence. Till today, we are adapting in ways to coexist with this vital resource, building and rebuilding our infrastructure, policies and behaviours, while also acknowledging that water's flow is often unpredictable—reminding us that no matter what we do, it will remain on the edges of our control.

In 2024, Europe experienced its most widespread flooding in over a decade, with nearly a third of the continent's river network exceeding high-flood thresholds. Among the hardest hit regions was Valencia in Spain where hundreds perished. In Central Europe, Storm Boris (September 2024) unleashed record rainfall, and scientists from the World Weather Attribution (WWA) group found that climate change made those downpours at least twice as likely and roughly 7 to 10 per cent more intense.[17] These events are part of a broader trend of escalating flood risks globally and underline how climate change-driven extremes are now hitting places that were long considered secure and prepared.

Flooding can also make climate change worse. When floodwaters cover plants and other organic material, they start to break down and release methane, a powerful greenhouse gas. Rebuilding after a flood also uses a lot of energy, and if this comes from fossil fuels, it directly contributes to emissions. Rebuilding requires materials like cement and steel that together account for well over 10 per cent of global CO_2 emissions as their production is extremely carbon-intensive. On top of that, floods can damage forests and other ecosystems that naturally absorb CO_2, thereby increasing it in the atmosphere. All of this compounds the climate crisis, and we find ourselves in yet another *vicious cycle*.

UNDER THREAT: EARTH, AIR, FIRE, WATER, SKY

For centuries, ancient cultures and philosophies have spoken of five fundamental elements: Earth, Air, Fire, Water and Sky. In Greek thought, these elements were believed to form the building blocks of all matter. In Indian philosophy, they are known as the *Pancha Mahabhuta* or the five great elements. Similar ideas can be found in Chinese, Japanese, Tibetan and other traditions across the world, each recognizing the core forces of nature. While climate change is rooted in science, it is hard not to notice how it touches each of these elements. It affects the Earth we live on, pollutes the air we breathe, threatens the water we drink, disturbs the skies we travel through, and fuels fires that burn down forests, villages and cities. Philosophers may have debated the meaning of these elements in one context or the other, but what is clear is how deeply we rely on them. If climate change is disrupting Earth, Air, Fire, Water and Sky, the very elements that make up life, it is the biggest existential threat we face today.

It is taking human lives every day. Climate change and human health are deeply intertwined, with escalating environmental disruptions intensifying deaths and health risks across the globe. Extreme heat waves, now more frequent and severe, have led to a 167 per cent increase in heat-related deaths among individuals over sixty-five since the 1990s, as per the 2024 report of *The Lancet Countdown on Health and Climate*.[18] Simultaneously, prolonged droughts compromise food production, exacerbating malnutrition and undernutrition, particularly in vulnerable regions. Poor air quality, intensified by climate-induced factors like wildfires and increased ozone levels, contributes to respiratory and cardiovascular diseases, placing additional strain on healthcare systems. Flooding events not only damage health infrastructure but also increase exposure to waterborne diseases. Furthermore, the thawing of permafrost and shrinking wildlife habitats heighten the risk of zoonotic diseases, as humans and animals come into closer contact, potentially triggering future pandemics. In 2016, a heat wave on the Yamal Peninsula in Siberia thawed the ground, exposing an infected reindeer carcass that had been buried for decades in permafrost.[19] The bacteria's dormant spores were released into the environment, triggering an anthrax outbreak that killed at least one child, hospitalized dozens and wiped out over 2,300 reindeer. Terrifying, really.

THE SOLUTION BEGINS WITH UNDERSTANDING

The year 2023, when the idea for this book first came to us, was the hottest ever recorded. As we continued writing into 2024, we realized that 2024 surpassed it as the hottest year on record. According to the WMO, 2024 became the first year in which global temperatures rose more than 1.5 degrees Celsius above pre-industrial levels, a threshold scientists have warned about for decades. The United Nations has described this sequence of record-breaking years as a triple whammy, a run of extreme heat that risks pushing parts of the climate system towards irreversible change.[20] By the time you hold this book in your hands, 2025 will also be confirmed as one of the hottest years humanity has ever lived through.

Today, atmospheric CO_2 levels are about 50 per cent higher than they were in the pre-industrial era, according to the WMO.[21] Arctic sea ice has been shrinking at an alarming pace, declining by nearly 13 per cent per decade since 1979 as reported by the National Snow and Ice Data Centre.[22] Scientists estimate that the world has already lost more than half of its living coral reef cover since the mid twentieth century, driven largely by warming oceans.[23] Satellite measurements from NASA show that global mean sea level has risen by more than 100 millimetres since 1993, and if the current rate continues, the world could see a further rise of roughly 160 to 170 millimetres by mid-century.[24] In some years, global economic losses from extreme weather have exceeded US$300 billion, and leading climate and health assessments warn that without significant action, climate change could contribute to many millions of additional deaths by 2050.

This chapter may have felt heavy, filled with the many challenges that climate change brings. But there is a reason we began here. To design inclusive and impactful solutions, we must first understand the complexity and interconnectedness of the problem. Without that, the solutions will never match the scale of the challenge. You will see this book shift from a story of crisis to one of transformation.

Climate change is real. It affects every part of our world today. It worsens existing problems and deepens vulnerabilities, turning challenges into multiplying threats. It can create *vicious cycles* that are difficult to break. These cycles can push families like Asha's, who already live on the margins, into poverty traps that become harder to escape over time.

There's something else that has started to touch every part of our lives: technology. It now impacts how we communicate, learn, work and access services, offering a powerful point of intervention. Small, well-designed tech changes can scale quickly and reach people who had remained disconnected from the digital world so far. Digital tools can amplify impact, streamline services, support real-time decisions and connect marginalized communities to vital resources. Framing technology as a central lever helps us design interventions that are efficient, inclusive and able to transform systems at scale.

Asha's family was left without a roof over their head after a conflict triggered by the drought and the subsequent food shortages. Climate change played an important role in creating those shortages. Such conflicts could intensify as climate change continues to strain food systems. Sophie received an alert on her phone, warning her to evacuate as wildfires swept through her town. The heatwave that fuelled those fires was quite possibly driven by climate change. This raises a deeper question. Could the very technologies that hold transformative potential to keep us safe, like AI-powered alert systems, also be vulnerable to the impacts of climate change themselves?

To reach that point, we need to unpack a few more important connections.

3

TECH ON THIN ICE

When Technology Gets Caught in the Climate Crossfire

Sophie and Asha both have smartphones. Sophie's house, which was destroyed in the fire, had a landline phone many years ago, before mobile phones made them redundant. She had never used one herself, but had seen photos of her father speaking on the landline phone in the living room, her mother standing beside him with a hand on his shoulder. Sophie's parents had opened an account with the local bank branch years ago. Sophie, though, had only heard stories about depositing and withdrawing money at the bank. Growing up, she had only ever seen her parents using a banking app on their phones.

Asha too had a smartphone, now charging from a battery whose cables ran haphazardly through their tent in a relief camp for refugees. But there was no signal. Telecom towers had been destroyed in the conflict. For her, the phone was a window to the world beyond her village. She missed her uncle, who had moved abroad ten years ago. He used to connect with them through a video call on her phone, smiling as he showed her the streets and sights of the city that he now called home. He had always been the ambitious one in the family, and word had spread among friends and relatives that he was doing well enough to potentially run for political office soon. When Asha's parents were growing up, their village had no bank, and neither electricity nor telephone lines had ever reached them. Their first connection to the outside world came through a mobile phone. Their first experience of banking was through a mobile app, a powerful example of how many in the developing world, or the Global South, are leapfrogging years of traditional infrastructure and jumping straight onto the digital bandwagon. But with the internet down and their region now in turmoil, they had to rely on borrowing money just to get by. Years of progress, undone by a complex set of factors that lie across both climate and conflict.

ONE GLITCH AWAY

Technology has become a crucial and powerful part of our everyday lives. We use it to communicate with friends and family, access information, navigate our surroundings, manage our schedules and conduct financial transactions. We buy, sell, build communities and fall in love, sometimes entirely online. Social media use has become nearly ubiquitous, engendering new forms of creator economies that rely solely on social capital to thrive.

The global influencer market size (yes, there is a market for this too) is expected to rise from USD 9.89 billion in 2024 to USD 34.79 billion by 2033.[1] From self-driving cars to Internet of Things (IoT) at home—that controls everything from the music we play to the lights in our living room (whatever happened to walking *just* a couple of steps to switch off the fan?)—everything can now be done with the click of a button, or a simple voice command.

In an increasingly interconnected world, our deep reliance on technology brings with it a significant vulnerability: even minor disruptions can have sweeping consequences. In 2024, a global outage was caused across all Windows systems when cybersecurity company, CrowdStrike, pushed out a faulty update, which led to widespread *blue screen of death* (BSOD) errors, halting operations for businesses and individuals.[2] You could see the BSOD staring at you from your office computers, from airport flight-information display screens and even self-checkout kiosks at supermarkets. Flights were grounded and hospital IT systems froze.

Meanwhile, recurring global outages of platforms like WhatsApp, Instagram and Facebook have revealed just how embedded these technologies are in modern life. While younger users, primarily Gen Z and millennials, may flood X (formerly Twitter) with frustration over disrupted social lives, the impact runs deeper. Small businessowners, many of whom rely on tools like WhatsApp Business for daily operations, suffer real economic losses during such blackouts. For them, a few hours of downtime can mean missed orders and a significant loss of revenue.

Just a few years earlier, in 2020, the world witnessed another stark reminder of how fragile our tech-reliant systems can be. Sparked by pandemic-related disruptions, surging demand for electronics, and supply-chain vulnerabilities, the world was suddenly confronted with a global semiconductor shortage. The crisis exposed the extent to which everything from smartphones and laptops to cars and medical devices depend on a steady flow of microchips. Entire automotive assembly lines came to a halt, tech-product launches were delayed, and prices for consumer electronics soared. The ripple effects of that crisis echoed beyond factories, serving as a wake-up call about the geopolitical, economic and social consequences of overdependence on a few concentrated supply sources in an increasingly digitized world.

WHEN CLIMATE HITS TECH

Modern technologies such as smartphones, data centres and artificial intelligence systems depend on a wide range of critical minerals. These minerals are essential inputs for semiconductors, batteries, wiring, magnets, optical components and communication networks. They include materials such as silicon, gallium and germanium for chips, lithium, cobalt and nickel for energy storage, copper and aluminium for electrical infrastructure, and rare earth elements for high performance magnets, precision optics and fibre optic systems. Their supply is often concentrated in a few locations, which makes them strategically important for both industry and national security. Even AI chips or GPUs, which sit quietly in massive data centres and perform large, complex calculations to power AI tools, depend on these materials. That includes the very tool—ChatGPT, Perplexity or DeepSeek—that you were probably overloading a few hours ago with endless questions.

Semiconductor manufacturing is a highly complex process requiring ultra-clean facilities, enormous water usage, and precise environmental controls. Prolonged droughts and water shortages caused by climate change can severely impede production, leading to delays and shortages of these vital components. Countries such as China, Myanmar, Australia, United States, Malaysia, Brazil and even Greenland where REEs are mined, or are becoming attractive destinations for their exploration, are becoming increasingly vulnerable to climate-related risks such as floods, droughts and extreme weather. These disruptions affect not just mining but also the global manufacturing hubs where semiconductors and related technologies are assembled. With more than 70 per cent of the world's semiconductor manufacturing concentrated in a single country, the global supply chain is highly exposed. Asha's country sat on a major transcontinental highway that once carried goods across borders, but now, with drought-induced food shortages fuelling local conflict, roads are blocked, the route is shut, trucks take long detours, and the semiconductors that power her phone reach their destination slower and costlier than ever.

Climate change has the potential to disrupt every technological device we use today, including AI systems that are becoming integral to modern society. Even the International Energy Agency (IEA) warns that critical technologies, including semiconductors, are vulnerable to shocks in the supply chain, and their demand is only expected to increase as the transition to clean energy gathers pace.[3] It is interesting to note that the technologies that can accelerate the clean energy transition—helping combat climate change—are also very prone to shocks that are brought about by climate change. Recent analysis highlights that climate-driven water stress poses a serious threat to the semiconductor supply chain. Approximately 40 per cent of existing chip plants, along with over 25 per cent of facilities under construction, and more than 40 per cent of those announced since 2021, are located in watersheds that are expected to face high or extremely high water stress by 2030–40 as per a report by the World Economic Forum.[4] PwC estimates that by 2035, climate-related water stress could threaten a third of global semiconductor supply.[5] This matters because chip fabrication is incredibly water-intensive, and disruptions in critical regions could ripple across the entire network of the world. A few months ago, when people went into a frenzy creating Ghibli-style images with ChatGPT, Sam Altman had to announce on Twitter that limits were being imposed on searches because the queries were 'melting GPUs'. These GPUs are located in data centres, which are critical infrastructures that are increasingly vulnerable to the impacts of climate change. These facilities require stable environmental conditions and substantial energy, particularly for cooling systems. However, rising global temperatures and extreme weather events pose significant threats to their operations. For instance, in July 2022, a severe heatwave in the United Kingdom led to the overheating and subsequent shutdown of Google Cloud and Oracle data centres in London, disrupting services to various customers. Data centres already consume massive amounts of electricity, and nearly 40 per cent of that energy goes towards cooling servers to prevent overheating, if estimates by the National Renewable Energy Laboratory (NREL) are to be believed.[6] As heatwaves intensify with climate change, these cooling demands increase even further. That means more power consumption, and in many cases, more greenhouse gas emissions. In essence, rising heat leads to higher energy use to stay cool, and that added demand contributes to even more warming. Once again, our *vicious cycle* is in full force.

IS THE HEAT ON MOORE'S LAW?

In 1965, Intel co-founder, Gordon Moore, observed that the number of transistors on a microchip doubles approximately every two years, while the cost of computing is halved. In essence, computing power grows exponentially, while its cost per unit keeps falling. This is known as Moore's Law, an iconic principle which has come to represent the broader trend associated with technological innovation.

For most people, this sounds abstract. While the law was not meant to be universal and applicable to everything, perhaps the spirit of it can be observed across other sectors: exponential improvements in technology with a decline in costs. Years ago, televisions were rare in villages in India. The wealthiest family in the village owned a large, boxy CRT set, and entire communities would gather under a tree to watch a sports match together. Fast forward to today, and homes have sleek OLED TVs, which are much more advanced, pack hundreds of features and can be connected via the internet to streaming services. This is what progress has looked like. Technology has grown faster, smaller and more affordable. From plasma screens to OLEDs, from smartphones to cloud computing, innovation has made powerful tools accessible to more people than ever before. The same principle—more performance for less cost—also applies across other fields. Lithium-ion batteries are more efficient; smaller battery sizes can pack more power and yet, their prices have dropped by nearly 90 per cent in the past decade.

But as we move towards the future, a number of threats are playing out in real time. The previous chapters touched upon some of them, and perhaps we can all agree that effects of climate change throw a spanner in just about every sector, complicating existing problems and compounding their impact. Floods, wildfires, droughts and extreme heat are no longer distant risks. They are here, and they are taking lives and disrupting global supply chains. In 2021, severe floods in Malaysia delayed chip production. In 2022, extreme drought in China led to power shortages and factory shutdowns. Data centres have been knocked offline due to heatwaves. Modern technology depends on a fragile, global web of production and integration. The lithium for your electric vehicle battery might be mined in Argentina, processed in China and assembled in India. If any link in that chain is disrupted, whether by climate events or geopolitical tensions (which themselves can be compounded by climate change), then the entire system feels the shock.

The bigger picture is this: exponential progress only works if the physical systems that support it can keep up. Our oversimplified, in-principle explanation of the spirit of Moore's Law does not hold if climate change disrupts the supply chains that technology depends on. And yes, that was a heavy phrase. It was meant to be. Consider it our version of a friendly disclaimer. We are not here to rewrite or reinterpret Moore's Law. That would risk a revolt from engineers, scientists and researchers. We are simply trying to explore what this iconic idea means for the rest of us, those outside silicon labs, living in a world where computing power keeps growing, even as the planet heats up and systems begin to strain.

AI IN THE CROSSFIRE

Looking at the broader trend behind Moore's Law, the idea that computing power rises exponentially even as costs fall, we see that AI pushes this trajectory forward at remarkable speed, thanks to leaps in hardware, algorithms and scale. However, this progress depends heavily on stable supply chains and resilient infrastructure, both of which are increasingly vulnerable to climate change. Without a steady supply of semiconductors, the rapid pace of AI development and deployment could significantly slow or even stall. Beyond production, data centres where AI computations happen are highly susceptible to heatwaves and extreme weather. When data centres go offline, AI services and other critical internet functions slow or stop. Additionally, fibre-optic networks—that link data centres worldwide, enabling instantaneous data flow and cloud computing—can be damaged by climate extremes like floods or storms, leading to infrastructure failures. Such connectivity disruptions can fragment AI services, reduce efficiency and hamper innovations reliant on real-time data. Disruptions to semiconductor supply, heat-induced data centre failures and damaged communication networks, could collectively slow AI's momentum and increase costs, challenging the usual trajectory of rapid technological progress.

So, can climate change disrupt the supply chains and infrastructure essential to AI and other advanced technologies? Absolutely. Can unchecked AI contribute to climate change as it demands more power and resources? Yes. And can AI also be a powerful tool in mitigating climate impacts? Without a doubt. This delicate balance places AI right in the crossfire—a perpetrator, a victim and a vital part of the solution in our changing world.

The conflict that hit Asha's village, which we now know sits along a major transcontinental highway, is one of the reasons trucks carrying tech-essential minerals now take longer, more complicated routes. That means delays, higher costs and rising prices by the time those materials reach the other end of the supply chain. This conflict did not happen out of nowhere. It was triggered by a drought. And as we have already seen, climate change can cause or worsen droughts. So, that brings us to a bigger question. *Can climate change lead to conflict?* And can that conflict then disrupt supply chains, slow down technological progress, and make it harder for governments to focus on climate action?

Could this become yet another *vicious cycle*?

4

THE GEOPOLITICS OF CLIMATE CHANGE

Is Climate Threatening the World Order?

VULNERABILITY AND THE SEEDS OF CONFLICT

Asha's village was located in a country which, until a few years ago, was known for its thriving agricultural produce. They did not just feed their own country, they exported to the world. Along with being a major agricultural hub, the region was also fairly rich in rare earth elements (REE). By now, you know why that matters, both for technology and for this book. It was this unique combination of agricultural richness and mineral wealth that led to a major transnational and transcontinental road passing directly through Asha's village. Village elders often told her that while water had always been scarce, they received just enough rainfall and had traditional conservation practices that supported healthy, flourishing crops. But then, things started to change. Rainfall became irregular. The past few years brought little to no rain at all. Prolonged dry spells led to drought-like conditions, and agriculture began to suffer. Scientists studying the region believed that climate change had a major role to play.

This story is not unlike what has been unfolding in Chile, a country that has endured drought-like conditions since 2010, and continues to grapple with increasingly scarce water resources as the dry period extends well into its second decade. Experts believe the causes lie in a mix of climate change and the overexploitation of agricultural land. Asha's country was part of a larger region that had once been crudely divided by colonizers from the West. Before the colonizers left, borders were drawn with rulers on paper maps, without thought for how communities lived, how resources were shared, or how the future might unfold. These lines were never about fairness. The colonizers had already taken what they wanted, and when they left, they left behind fragmented nations.

That history meant civil unrest was always simmering beneath the surface. When you divide land based on arbitrary lines, you ignore how people access water, food and land. And when climate change pushes resources to the brink, those old lines turn into real conflicts. As food supplies dwindled with the prolonged drought, tensions rose. Armed groups from within the country and across the border entered into a complicated and violent struggle. The transcontinental road that once brought prosperity was now blocked. Global supply chains began to feel the strain. The spread of conflict threatened a world that runs on fragile and interconnected systems of trade and diplomacy. Some of the same countries that had once drawn those borders decided they needed to intervene. But that intervention compounded problems rather than solving them.

Sophie comes from one of those countries that decided to get involved. She knew, in vague terms, what her nation had done in the past, though that part of history had never been covered in her textbooks. The fighter jets that roared through the skies in the first chapter, the ones Asha looked up at, had taken off from Sophie's country. But right now, neither Asha nor Sophie had time to reflect on the weight of that history.

CLIMATE CHANGE, SCARCITY AND RESOURCE COMPETITION

As we explored in the previous chapter, the prospect of a world brought to its knees by a breakdown in technological infrastructure, potentially driven or worsened by climate change, is already alarming. But now, an even graver question looms. Based on what Asha's country is currently witnessing, we must ask whether climate change could also become the catalyst for future conflicts, even wars. As the planet warms, it could intensify competition over vital resources such as food, water, energy, and critical materials ranging from REEs to essential metals like copper.

A study[1] carried out by the International Growth Centre found that a *one standard deviation increase in temperature may cause a 10.8 per cent average increase in conflict incidence and a 16.2 per cent average increase in the violent crime rate.*

Imagine two people fighting on a crowded street over a single parking slot. There is limited availability of parking spots, but everyone wants to park their vehicle. Another scenario, all too familiar from disaster movies, is people breaking into department stores as soon as disaster hits, to get their hands on the limited resources that are available. They most often do not seem to be doing this out of malice (although some do), but out of desperation. However, when everyone starts doing that, it leads to an even more complicated situation where some actors see the desperation and limited supplies as an opportunity to start hoarding, creating a black market of sorts. The thin line between civility and chaos frays fast when survival enters the equation.

But does it happen only in the movies? If you're reading this, you already know the answer to that question. In the early days of the COVID-19 pandemic, entire aisles of supermarkets were stripped bare when lockdown measures across the world were announced, with people hoarding everything from toilet paper to ready-to-eat meals and canned goods. Fistfights broke out. It was not just about the supply—there was enough of it. But even the mere *anticipation* of a potential shortage and the fear of going without essential items was triggering anxiety and causing chaos. There was also the fuel crisis in the United Kingdom in 2021. Just a few warnings about potential shortages led to long queues at gas stations, panic buying and even altercations at the pump. The fuel was still flowing, but the *mere perception of scarcity triggered chaos.*

Now let us put this into our climate change frame. Imagine changing weather patterns that lead to a prolonged dry spell in a region, which has a key river passing through multiple countries, not all of which may share particularly warm relationships with each other. As the flow of the river reduces, drinking water supplies will deplete, crops may perish, and in the absence of food and water, ripple effects will be felt across the entire economy. Countries which are uphill and have dams may try to use them to store more water, whereas the countries downhill will demand more water to be released to them.

Similarly, in situations of drought, local reservoirs dry up, groundwater runs low and irrigation systems falter. In agrarian economies, entire harvests may fail, pushing food prices up and forcing governments to impose export bans or hoard supplies. Neighbouring countries dependent on those imports could react with diplomatic pressure, or perhaps through trade wars or resource grabs masked as economic retaliation. In the worst case, it may just lead to an outright conflict. This is what is happening in Asha's village.

Now consider energy. A region reliant on hydroelectricity from melting glaciers or flowing rivers finds itself in trouble as snowpacks shrink and rivers run dry earlier each year. Power shortages become common, and countries begin to compete for remaining sources of electricity, either by trying to buy out energy from smaller neighbours or by investing heavily in strategic infrastructure across borders, stoking contestation.

Take critical minerals such as lithium, cobalt and rare earth elements (REE), which are essential for green technologies. As extreme weather events, water shortages and ecosystem collapses make mining and supply-chain operations more difficult, supply begins to choke, fuelling strategic competition. Though not yet at crisis levels everywhere, early signs of these dynamics are already playing out in various parts of the world

In Nigeria, for instance, climate change is exacerbating resource scarcity between farmers and herders.[2] The country is facing climate-related shocks that have triggered displacement, with herding communities forced to leave their traditional lands in search of better grazing areas, putting them in competition with settled farmers. These tensions over dwindling resources are further fuelling conflict, as competition for water and fertile land intensifies due to the changing climate. This situation is a growing concern for both humanitarian and security issues in the region.

The Colorado River, a vital water source for both the United States and Mexico, is facing significant challenges due to climate change, which has led to reduced water flow. In the United States, the Colorado River and its water-sharing arrangement with Mexico are under increasing strain as drought and long-term water scarcity limit available flow. In March 2025, the United States Department of State announced that the U.S. would for the first time deny a non-treaty request by Mexico for a special delivery of Colorado River water to Tijuana, citing Mexico's repeated shortfalls in meeting its commitments under the 1944 Water Treaty. On the other hand, Mexico has stated that severe drought and water shortages have constrained its ability to meet treaty-mandated deliveries, and that the issue is being addressed through the International Boundary and Water Commission.[3] Both countries depend on the river for drinking water, agriculture and energy production, and the ongoing dispute highlights how climate-induced water scarcity is escalating geopolitical tensions over shared resources.

In the Himalayas in South Asia, glacial melt is accelerating due to rising global temperatures. An alarming report published by the International Centre for Integrated Mountain Development, a regional intergovernmental agency, finds that glaciers in the Hindu Kush Himalayas could lose up to 80 per cent of their current volume by the end of the century if our emissions trajectory remains the same. The ice and snow in the Hindu Kush Himalayas serve as a crucial water source for 12 rivers that flow through 16 countries in Asia, supplying freshwater and essential ecosystem services to 240 million people in the mountains and an additional 1.65 billion people downstream. Under a high-emission scenario, the snow cover could fall by a quarter,[4] which could affect water flows to major rivers across Central and South Asia, including across countries like Nepal, Bhutan, India and China, where water-related disputes are already a component of complicated geopolitical equations. Any reduction in glacial mass may potentially affect river flows, agriculture and hydropower generation, posing significant challenges to regional water security.

SYSTEMIC RISKS AND THE GEOPOLITICS OF CLIMATE CHANGE

According to a Stanford University-led study published in *Nature*, in a scenario with 4 degrees Celsius of warming, *if societies fail to significantly reduce emissions of heat-trapping gases, the influence of climate on conflicts would increase more than fivefold, raising the chance of a substantial rise in conflict risk to 26 per cent.*[5] Even in a scenario of 2 degrees Celsius of warming beyond preindustrial levels, which aligns with the stated goal of the Paris Climate Agreement, the influence of climate on conflicts would more than double, increasing the likelihood of conflict by 13 per cent.

As the global transition to green energy accelerates due to climate change, the demand for minerals like lithium and REEs has surged. As per a report by the World Bank,[6] the production of minerals like graphite, lithium and cobalt could surge by nearly 500 per cent by 2050 to meet the increasing demand for clean energy technologies. It is projected that over 3 billion tons of minerals and metals will be necessary to deploy wind, solar and geothermal power, along with energy storage, to achieve a future where global warming is below 2 degrees Celsius.

Bolivia, with some of the world's largest lithium reserves, and Africa, rich in rare earth minerals, have become central to this race. Lithium is crucial for electric vehicle batteries and renewable energy storage, while rare earth minerals are vital for electronics and military systems. Countries like China and the United States are vying for control of these resources, with China expanding its influence through investments in mining operations across Latin America and Africa. This competition is intensifying geopolitical rivalries, as nations navigate the economic and security implications of securing access to these critical materials for the green energy transition. However, these regions themselves are also vulnerable to climate impacts, which could disrupt mining operations.

According to a report from the International Energy Agency,[7] between 2020 and 2024, the growth in refining critical minerals was heavily concentrated. Moreover, the average market share of top 3 nations into refining minerals crucial for the energy transition stood at almost 86 per cent in 2024. Of this, over 90 per cent of supply growth came from a single nation—Indonesia for Nickel and China for Cobalt, Graphite and REE. This high level of supply concentration poses a significant risk to the speed of energy transitions, as it makes supply chains and routes more vulnerable to disruptions, whether caused by extreme weather, trade disputes or geopolitical tensions.

It is worth noting that our *vicious cycles* have come back around. While the growing demand for climate action or the global transition to net zero is accelerating, climate change threatens the very areas where the supply of these essential minerals is concentrated. As the supply gets hit, global action on climate becomes stunted, accentuating the effects of climate change, and hitting even more reserves.

While countries play a crucial role in mitigating and managing the challenges that climate change can impose on critical supply chains, whether food, water, energy or minerals, the highly global nature of these supply chains requires strong regional and global cooperation. It is particularly important to note that climate-induced extreme weather events will be especially devastating for developing countries, as many of these nations have socioeconomically vulnerable populations and are located in areas more prone to climate risks. According to the IPCC's Sixth Assessment Report,[8] between 3.3 and 3.6 billion people live in settings that are highly susceptible to climate-related disruptions. These vulnerability hotspots are primarily located in small island developing states, the Arctic, South Asia, Central and South America, and large parts of sub-Saharan Africa. In these regions, poverty, conflict, weak governance and limited access to essential services severely constrain adaptive capacity, compounding the risks. The resulting humanitarian crises threaten to displace millions. Alarmingly, by 2030, drought alone could place an estimated 700 million people at risk of displacement.[9]

Broadly speaking, a large share of the world's newly mined critical minerals comes from countries in the Global South: cobalt from the Democratic Republic of Congo (DRC), nickel from Indonesia, and lithium from Argentina and Chile. But it is equally important to recognize that the value-chain does not stop there: much of the refining and manufacturing remains concentrated in a handful of other nations. This double-layer concentration means supply chains remain highly vulnerable, even as the Global South provides the raw materials. These countries face financial constraints as they must balance the challenge of lifting people out of poverty with investing in climate change mitigation and adaptation. This creates a *vicious cycle* again: investments in socioeconomic development are often undone by climate impacts, requiring funds to be redirected from climate adaptation and mitigation efforts to repair the damage caused by extreme weather events. As a result, there are fewer resources available for climate action, perpetuating the cycle. Climate-triggered conflict only acts as an accelerant in deepening this *vicious cycle*.

CLIMATE MULTILATERALISM

Since 2018, the UN's Climate Security Mechanism,[10] a joint initiative between the Department for Political and Peacebuilding Affairs, the United Nations Environment Programme, and the United Nations Development Programme, has provided multidisciplinary support to member states, regional organizations, and UN entities to better understand the links between climate, peace and security. The World Bank launched a new Digital Transformation Vice Presidency in January 2024, aimed at working with governments in developing countries to build the foundations for digital economies and societies.[11] This is crucial, as increasing digitization and deploying frontier technologies like AI can improve infrastructure planning, predict extreme weather events, model climate risks, manage agricultural productivity, optimize water flows, and build resilient infrastructure and supply chains.

India, in particular, is taking a leading role in amplifying the voice of the Global South. The Voice of the Global South Summit,[12] convened by India in 2023, brought together over 125 countries to discuss issues of climate justice, equitable development and global solidarity. Beyond the summit, India has also championed platforms like the International Solar Alliance, co-founded with France, which aims to mobilize over a trillion dollars in investment towards solar energy by 2030. Similarly, India's leadership in initiatives like the Coalition for Disaster Resilient Infrastructure and the Global Biofuels Alliance underscores its commitment to fostering resilient, sustainable development pathways for the Global South. These efforts reflect not only a commitment to climate action, but also a recalibration of global leadership where the concerns, aspirations and agency of the Global South are placed at the centre of global decision-making.

However, we have seen that multilateral institutions have, at times, proven inadequate in managing the complex challenges of the 21st century. When many of these institutions were established, the challenges they faced were less multifaceted and complex. They were not designed to tackle issues such as those induced by the COVID-19 pandemic or the widespread havoc being caused by climate change.

While they have slowly started adapting to the complex challenges of climate change, their efforts must deepen radically and quickly, especially in mobilizing the scale of finance needed. Without significant and sustained financial support, it is unrealistic to expect emerging and vulnerable economies to effectively face the growing climate crisis. Climate adaptation and mitigation require resources far beyond what many developing nations can provide alone. As a global community, the key question becomes who should bear the financial burden of combating climate change. Fair and effective climate finance must reflect not only collective responsibility, but also historic and ongoing contributions. Understanding these responsibilities and how they translate into action is critical as we look ahead.

5

RETHINKING CLIMATE EQUITY

Humanity's Shared but Differentiated Burden

It was day three for Sophie at the makeshift relief centre set up in the community stadium. When she used to come here to play or watch sports matches, she had never imagined that one day it would be filled with tents, medical stretchers and relief counters, and even less that she would be there herself. That morning, she had witnessed some heartbreaking scenes. Families were frantically showing photos of missing loved ones, asking anyone if they had seen them. Many people were still unaccounted for in the aftermath of the wildfire.

After consuming their town, the wildfire had spread to the next one, and firefighters were struggling to bring it under control. The television screens, which once showed game scores and advertisements for clothing brands, were now tuned in to the news. The wildfire dominated every channel. As the news report ended, the anchors moved briefly to international updates. A wide range of stories appeared: the concerning rate of deforestation in the Amazon, an ongoing tariff war, the winner of Wimbledon, and their country's president announcing a decision to send fighter jets to a nation thousands of miles away, in the name of global stability and resource sufficiency.

Someone near Sophie changed the channel, muttering, 'Let us first put our own country in order', and tuned in to another channel offering more coverage of the wildfire. Curious about what had led to her President's intervention in that faraway nation (Asha's country), Sophie pulled out her phone and opened Perplexity, one of the many AI apps she used, alongside ChatGPT, DeepSeek and others. She asked about the country and the war. As she read how climate change had led to a drought, which had then triggered conflict and instability, she came across a speech made by the former President of Asha's country, at the United Nations. Standing at the podium, the President delivered a powerful message about her country's role in the climate crisis. She emphasized that nations like hers were not responsible for causing climate change, yet are now suffering its harshest impacts. She described her country's limited financial capacity to confront these challenges, calling on the international community to step up urgently with funding, technology, and meaningful support, rather than expecting vulnerable nations to shoulder the burden alone. Despite the many pressing concerns her country faces, the President highlighted the remarkable progress they had made in renewable energy, outpacing the achievements of many developed countries. She underlined the principle of 'Common But Differentiated Responsibility', urging a global response rooted in equity and shared commitment. She was the President of Asha's country, and she was calling on the world for help.

Sophie's President, meanwhile, dismissed climate change as a hoax at the same forum.

UNDERSTANDING RESPONSIBILITY

Climate change is a global challenge borne by all of humanity, but the burden of responsibility is unevenly spread. This imbalance is captured by the internationally recognized principle of Common But Differentiated Responsibilities (CBDR), formalized in the United Nations Framework Convention on Climate Change (UNFCCC) at the 1992 Rio Earth Summit. The principle acknowledges the reality that while all countries share the obligation to protect the climate, not all have contributed equally to the problem nor do they possess equal capacity to address it.

Historically, industrialized countries advanced by burning vast amounts of fossil fuels such as coal, oil and gas, releasing massive amounts of greenhouse gases that accumulated in the atmosphere. These nations enjoyed the benefits of early industrialization, which powered economic growth and improved their living standards. This is also why they are 'developed', today. In contrast, many developing countries are now striving to elevate the quality of life for billions of people, but face a significant constraint: the planet's atmospheric carbon space, or the capacity of the atmosphere to safely absorb CO_2 without triggering catastrophic warming, is finite and has been largely consumed by those who industrialized first.

Imagine the Earth's atmosphere as a global credit card with a fixed spending limit. This spending limit is the amount of carbon space that the world can occupy. Think of it as our collective carbon (or emissions) budget. Developed nations have swiped large portions of this limit during their industrial growth, leaving less room for developing nations to use in their pursuit of economic progress. This raises a crucial question: how can developing countries grow sustainably when the 'carbon budget' available to them is severely limited? The ethical and practical challenge of carbon equity lies at the heart of climate diplomacy and global cooperation.

THE CARBON FOOTPRINT DIVIDE

Carbon emissions demonstrate stark global disparities, both historically and in the present. Currently, developed countries have consumed approximately 82 per cent of the world's historical carbon space.[1] Developing nations, with much larger populations, are left with a smaller share to emit while they seek growth and prosperity. India, for instance, despite being the world's most populous country has taken up approximately only 1.5 per cent of the remaining carbon space.[2]

India's per capita carbon emissions in 2024 stood at around 2.20 metric tons of CO_2 annually, among the lowest for major economies. Brazil's per capita emissions are slightly higher at 2.28 metric tons. Ghana, a smaller economy of around 34 million people, emits as low as 0.61 metric tons per person.

Contrast this with wealthier nations:

The United States emits about 14.2 metric tons of CO_2 per person each year, roughly seven times India's level despite having less than one-fourth of India's population. Norway emits a little over 6.67 metric tons per person, around eleven times higher than Ghana's rate.[3]

When we look at cumulative carbon emissions—the total CO_2 emitted since industrialization—the global disparities become stark. The United States alone has released over 500 billion metric-tons of CO_2, accounting for about 20 per cent of the global total.[4] In comparison, despite its population of over 1.4 billion, India's historical contribution to global emissions remains only around 4 per cent.[5]

This historical context underscores a profound asymmetry: developed nations benefited from a prolonged phase of carbon-intensive growth with limited regard for climate impacts (which, to be fair, were largely unknown at that time). In contrast, developing countries today face pressure to pursue low-carbon growth pathways under significantly constrained carbon (and indeed, monetary) budgets.

THE TRIPLE CHALLENGE FOR DEVELOPING COUNTRIES

Developing nations bear a triple burden in the global climate landscape. Firstly, there is the pressure to rapidly decarbonize, a demand that the developed world never faced during its own period of industrialization. While this expectation is understandable, given the accelerating pace of climate change, it must be paired with assured access to adequate finance and cutting-edge technologies for these countries.

Secondly, developing countries must continue expanding their economies to lift millions, and in some cases billions, out of poverty while building industry, infrastructure and resilient, inclusive societies that can sustain future growth. These essential tasks demand significant energy and investment, and require difficult, carefully calibrated choices: should each available dollar go towards eradicating absolute poverty or be used for transitioning away from coal and other carbon-intensive activities?

Thirdly, these countries already bear the harshest impacts of climate change including rising temperatures, extreme weather, floods, droughts and displacement, despite having contributed little to the emissions that drive these crises. According to the World Meteorological Organization, over the past fifty years, weather, climate and water-related disasters have caused more than two million deaths worldwide, with over 90 per cent of these occurring in the developing world.[6] With their financial resources already stretched thin, imagine if that one dollar was originally meant to lift someone out of poverty, or if the plan was to split that dollar evenly between decarbonization and poverty alleviation. The reality now forces them to spend one and a half dollars simply to rebuild after floods or wildfires destroy original infrastructure. Recovery is tough, and safety nets are often limited or nonexistent, deepening vulnerability and prolonging hardship.

WOEFUL CLIMATE FINANCE DELIVERY

The principle of CBDR remains central to international climate governance. The Paris Agreement of 2015 reaffirmed this principle, emphasizing that while all nations must act on climate change, those with greater historical emissions bear a greater responsibility. It committed developed countries to mobilize $100 billion annually by 2020 to support developing countries in their efforts to adapt, mitigate and manage climate risks.[7] However, progress on climate finance has been uneven, with much of the funding arriving as loans rather than grants, which adds to the debt burden of vulnerable nations. The establishment of the Fund for Responding to Loss and Damage at COP27 in 2022 marked an important political milestone: the first dedicated financial mechanism to support vulnerable developing countries facing unavoidable climate impacts.[8] While symbolizing progress towards climate justice, the Fund remains in early stages: at COP28 in 2023, parties agreed to its governing instrument and cut a deal to operationalize it, at COP29 in 2024 the fund's interim secretariat host was confirmed and as of September 2024, pledges surpassed USD 700 million.[9] At COP30, there was a renewed call to scale up climate finance for developing countries.[10] The critical question remains when these pledges, announcements and commitments will translate into scaled and timely finance, transparent governance and full access for the countries and communities most affected.

The scale of climate finance needed by developing countries far exceeds commitments so far. Estimates suggest that developing nations will need around US $1.1 trillion annually from 2025, rising to roughly US $1.8 trillion by 2030, to fund mitigation, adaptation and unavoidable climate-impacts like loss and damage.[11] At COP29 developed countries agreed to lead in mobilizing at least US $300 billion per year by 2035, and to work together to scale up flows to a total of US $1.3 trillion per year from all sources.[12] By contrast, current annual climate-finance flows from developed to developing countries remain in the order of US $115-120 billion, well below what is required.[13] These gaps underscore the urgent need to scale up climate finance several-fold if developing countries are to pursue sustainable development pathways while effectively addressing climate change.

The inequity is undeniable: those who have contributed least to climate change endure its most severe consequences. This profound injustice underscores the urgent need for equity in climate action, finance and governance.

Yet, within every challenge lies a powerful opportunity for transformative change.

6

CHALLENGES AS OPPORTUNITIES

Leapfrogging to a Sustainable Future

THE CHALLENGE

As a quick recap, we have seen that the global landscape is as fragile as it is uncertain, with the world grappling with a cascading series of crises, from trade wars and geopolitical flashpoints to stagnating economic growth and deepening climate pressures. Against this backdrop, climate change emerges as the most pervasive and complex threat, which also combines with existing threats and vulnerabilities and amplifies them. No sector remains insulated from its impact. It can trigger conflicts or exacerbate existing ones, it can lead to food shortages or worsen them, it can disrupt global supply chains, and it can even impact the technology that we use today, and worse still, it can create or amplify a *vicious cycle* where all of these things come together to overwhelm countries: for instance, climate disasters obliterate hard-won progress, forcing vulnerable countries to redirect scarce financial resources from proactive climate mitigation and adaptation efforts to costly recovery and rebuilding. Those already living on the economic margins are pushed deeper into poverty traps, with limited safety nets to soften the blow. The contrast in emissions and responsibility is striking. Early industrialized nations accounted for the majority of historical emissions, while developing countries continue to deal with balancing poverty reduction and climate commitments under significant resource constraints. Populations least responsible for historic emissions bear the harshest impacts, deepening global inequities.

The climate crisis, despite its gravity, offers a pivotal moment for developing countries.

OPPORTUNITY ONE: CLIMATE LEADERSHIP IN A SHIFTING GLOBAL LANDSCAPE

The developing world can play a leadership role in the global climate conversation and shaping international negotiations amid a fractured world order and evolving multilateral institutions *(that are currently undergoing successive waves of reorganization to better adapt to and address the complex and interconnected challenges of the contemporary world)*.

For instance, Indonesia's 2022 G20 Presidency linked green recovery to the pandemic response by advocating the shift of fossil-fuel subsidies towards renewables and urging equitable access to climate finance and technology.[1] India's 2023 Presidency championed just and inclusive energy transitions, calling for urgent reforms in institutions like the World Bank and IMF while pushing for clean-energy investment, global technology cooperation and resilient infrastructure.[2] India also launched the Global Biofuels Alliance and proposed a Green Hydrogen Innovation Centre, and a landmark achievement of its Presidency was the inclusion of the African Union as a permanent member of the G20.

Brazil's 2024 Presidency established the Task Force on a Global Mobilization against Climate Change to align finance with the Paris Agreement's 1.5-degree target and strengthen climate-finance commitments.[3] This is also the first time in G20 history that four emerging economies have held the Presidency in succession: Indonesia, India, Brazil and South Africa. South Africa, under its 2025 Presidency with the theme 'Solidarity, Equality, Sustainability', secured a Leaders' Declaration that reaffirmed support for vulnerable countries and emphasized affordable, predictable climate-finance.[4]

This leadership marks a shift from passive participation to active agenda-setting. But the momentum must not be lost.

OPPORTUNITY TWO: LEAPFROGGING TO GREEN GROWTH AND SUSTAINABLE DEVELOPMENT

While the industrialized world advanced through diverse stages of carbon-intensive development, the Global South now faces unprecedented pressures to decarbonize rapidly amidst persistent developmental needs. However, they are uniquely positioned to leapfrog traditional pollute-then-clean pathways by harnessing emerging green technologies in a massive way. (Remember how Asha's parents got their first experience of banking directly on a smartphone? This is what leapfrogging several stages of development entails.)

India, for example, is leading a green hydrogen revolution through its National Green Hydrogen Mission, targeting 5 million metric tons of green hydrogen by 2030 as a means to decarbonize heavy industry and transportation.[5] The country achieved its Nationally

Determined Contribution (NDC) target of securing 50 per cent of installed electricity capacity from renewables ahead of schedule, a milestone not yet surpassed by many developed countries.[6] With 85 per cent of its electricity coming from renewable sources, Brazil is at the forefront of Latin America's renewable energy expansion, rapidly increasing wind and solar capacity to strengthen energy security and enable sustainable economic development.[7] Meanwhile, Kenya exemplifies Africa's climate adaptation efforts by deploying innovative water management technologies, including solar-powered pumps and more efficient irrigation systems, augmented by digital climate services to bolster agricultural resilience amid recurrent droughts.

> *Thus, while it is true that the developing countries have been the lowest emitters and yet they have the highest vulnerability to impacts of climate change, they also have the greatest opportunity— and that is to industrialize without the need to carbonize.*

Unlike the Industrial Revolution, when viable low-carbon technologies were largely unavailable, today's advanced climate technologies enable these nations to bypass carbon-intensive phases and adopt sustainable pathways from the outset. India's green hydrogen mission, for instance, serves as a prime example of progress that is already underway independent of external climate finance, signalling proactive leadership. However, broader financing and tech-transfer remains critical for scaling solutions for the developing world.

The developing countries are already leading the global fight against climate change, and now, there is evidence to back it up. For the first time, electricity generated by renewable energy overtook coal and gas in the first half of 2025.[8] The growth in renewables outpaced the growth in electricity demand, even bringing down, albeit slightly, the use of coal and gas. It is also not surprising that this growth came from developing countries. India, China, the EU and the US continue to dominate the global energy landscape, and yet the use of coal and gas has fallen in India and China on the back of massive expansion in renewables. The EU and the US witnessed an increase in the use of coal and gas.

Now, let us dive into some of the most exciting advanced climate technologies that are already helping to speed up the green transition. These innovations have the power to unlock truly sustainable futures, so it's worth discovering what they can achieve!

7

ON THE CUTTING EDGE

Climate Technologies Enabling Transition and Shaping a Greener Tomorrow

Technologies that accelerate the world's transition to a net zero future not only exist, but are being developed, evolved and scaled at an unprecedented pace. Newer technologies are being developed in research labs, being pioneered by startups, and implemented at scale by companies and countries.

Asha and Sophie, although separated by thousands of miles, both loved going to their school. Sitting in the stadium, which had been converted into a temporary shelter after the wildfire tore through her town, Sophie reached for her bag. It was one of the few things she had managed to save. She looked through it to find something to read, hoping to take her mind off the frightening visuals of people screaming and scrambling during the wildfire. She found her history book and turned to the chapter they had last been studying in class; it was about the Industrial Revolution. The stadium was illuminated by power-guzzling incandescent floodlights, and the massive air conditioning system generated a constant hum, the aluminium ducts vibrating as cool air swept through them and exited through the vents. All of this powered by a grid which was drawing 60 per cent of its power from coal.

Asha was in a tent with her mother, much of their village had now been destroyed by the conflict. She had saved some supplies from the relief material she had queued up for the day before, and was preparing to have dinner. She missed her friends and she missed school. She remembered how her last science class had been about someone who had invented the light bulb. She looked up at a lone LED bulb hanging from the hook of the tarpaulin that made up the tent. The cables from hundreds of these tents ran to batteries, which, in turn, had been hooked up to a few solar panels, adjusted shabbily over stones and bricks in the direction of the sun, which would now rise tomorrow. Before the conflict, these panels had been on their rooftops, and villagers had been able to go back and salvage some of them.

Many of us may recall from school history or science lessons that James Watt was instrumental in greatly improving the steam engine, setting the wheels of industrial revolution in motion. (Interestingly, Hero of Alexandria demonstrated, in 1st century CE, that steam could be used to produce motion.) Edison came up with the light bulb and electric systems that became the means to power houses and factories. Michael Faraday demonstrated that electricity could be used to produce motion. Alexander Graham Bell invented the telephone. Perhaps, fewer people may know that the French inventor, Girard, and his son, experimented with machines to capture energy from the waves of the sea, or that Tim Berners-Lee gave us the World Wide Web, or the fact that Enrico Fermi and Leo Szilard pioneered nuclear power, unless of course, you are familiar with these fields.

And who gave the world the first viable green hydrogen electrolyzer? Who thought about kerfless wafering? Do we know who came up with Direct Air Capture (DAC) for carbon, or biochar for carbon sequestration? What about Gravity Energy Storage Systems (GESS) or CO_2-Plume Geothermal (CPG) systems? Who figured out that injecting ethanol into fuel could make such fuel less polluting? And what about the Solar Thermal Electrochemical Photo (STEP) process?

If these questions do not immediately bring names to mind, that is hardly surprising.

There are likely thousands more such innovations out there, and for many of them, you may not find a single credited name. This is not because they lack brilliance, but because they are the product of something even greater: decades of unrelenting research, cross-disciplinary collaborations and breakthroughs, and the convergence of multiple technologies all advancing in parallel, at a pace the past could never imagine.

If you do know the answer though, you may be someone who works closely with these subjects. Maybe you are a scientist, an engineer, a researcher, a policy analyst or someone with a sharp curiosity about how the world works. And perhaps, for asking what might seem like a sacrilegious question, you are already contemplating an open letter (which we will gladly read, preferably over a lawsuit) chastising us for doubting your intellectual instincts.

But before you sharpen your pen, pause.

This book is not written only for experts. It is written for everyone who inhabits this rapidly changing world. The general population is not one homogenous audience: it includes doctors, teachers, artisans, lawyers, historians, designers, builders, farmers, economists, bankers and thousands of other professionals who keep societies running. In their own ways, each of them enables, uses, questions or adapts to the technologies that shape our lives.

All fields matter. All forms of knowledge matter. And all people matter in the story of our shared future.

This book is for Sophie, and it is for Asha. It is for anyone who wonders what tomorrow might look like, and how to make it better. We hope for a world where children like Asha have a safe home, are protected from the fury of extreme weather, have access to good healthcare, nutritious food and clean water, can go to school, use a smartphone, and pick up and read any book they please.

The point of the original question was simple: the names of these inventors are not household knowledge. We no longer react with awe when we hear about breakthrough technologies, not because they are any less remarkable, but because innovation is now happening everywhere, all the time, and at a pace that is difficult to keep up with. The momentum is global; the scale is massive and the volume of change unfolding around us is extraordinary.

So, with that, let us dive into some of the many (read: a whole lot of) frontiers powering the transition to a cleaner, greener future.

If you are already well-versed in renewable energy, or if you would rather not pause for brief explanations and examples of a few recent innovations, feel free to skip ahead to the final paragraph of this chapter. The essentials will be right here, should you wish to revisit.

FRONTIERS OF RENEWABLE POWER

Power can be generated using coal, gas and diesel, but green and renewable power can be generated by harnessing elements such as the sun (solar), water (hydro) and wind. Solar power generates electricity using solar panels which have photovoltaic cells, and these cells convert sunlight into electricity. The wind-turbines, which now dot vast expanses of open areas such as hills, meadows and coastal plains in a number of countries, produce energy when wind spins turbines, which drive generators and electricity is produced. Both are clean and are becoming increasingly affordable. In India, electricity generated from solar power can be as low as US $0.03 per kilowatt-hour, a rate that marks the rapidly falling costs of renewables.

Within the realm of solar, innovations are unlocking greater efficiencies and cost reductions. For instance, perovskite solar cells can be more efficient, have lower production costs and can be applied to surfaces including glass and windows, converting them into potential sources of energy. While they do have their own challenges, after decades of lab research, manufacturers are shifting towards pilot-scale production, with hopes that commercial adoption may begin before 2030. Kerfless wafer technology is another significant advancement. It reduces waste during production of silicon (which makes up the solar panel) and can bring down production costs, while also lowering the carbon footprint of the production process itself.

We also have promising technologies such as the Solar Thermal Electrochemical Photo process, or STEP, which uses sunlight in two complementary ways. It generates electricity that powers electrolysis and at the same time provides direct heat to the reaction, which lowers the amount of electricity needed. STEP is unusual because it can cut emissions in some of the most energy-intensive industries such as hydrogen production and metal-making where high temperatures and fossil fuels are normally unavoidable. One of its most striking applications is a process that takes CO_2 and converts it into solid carbon such as carbon black. In doing so, it could potentially remove carbon from the atmosphere and turn it into a form that is stable, useful and does not pollute.

Yet another important source of renewable energy is hydropower. This form of energy harnesses the natural movement of water, typically from rivers, dams or waterfalls, as it flows over turbines and spins them. The mechanical motion of the spinning turbines is then converted into electrical energy through generators. Hydropower is one of the oldest and most widely used renewable energy sources worldwide, valued for its reliability, efficiency and ability to provide large-scale electricity generation.

When you walk along a beach and watch the steady rise and fall of the tides or hear the crashing of waves, you may be witnessing one of nature's most constant forces. Today, this constant and predictable force can also be used to generate electricity. Technologies like floating tidal turbines and oscillating water columns are at the forefront of this sector. Think of floating tidal turbines as underwater windmills that rotate with tidal movements and generate electricity. Oscillating water columns that tap energy from the oscillation of waves are rather fascinating. As waves rise and fall, they push air through a chamber, and that turns a turbine generating electricity.

If you thought wind turbines that generate energy were only deployed on land, you are wrong. Wind energy is increasingly being deployed at sea to capture stronger, more consistent winds than on land. These 'offshore' wind farms are groups of turbines installed in coastal waters that generate large amounts of renewable electricity. Compared to onshore wind, offshore farms benefit from higher and steadier winds, making them key to the clean energy transition. Floating offshore wind farms use turbines mounted on floating platforms anchored to the seabed. This allows them to be installed in much deeper waters where fixed foundations are not possible. In today's world where nature's power pushes our limits, 'typhoon class wind turbines' are not just withstanding this force but also capturing it to generate energy.

China's OceanX floating wind turbine is a remarkable example of such innovation. It has two turbines on a floating platform anchored at a single point, allowing it to face winds directly and maintain stability. OceanX stood firm in typhoon wind speeds between 133 and 152 kilometres per hour. Separately, 47 Goldwind turbines off Guangdong coast withstood 161 km per hour winds during Typhoon Yagi, producing 2.1 gigawatt hours in nine hours, equivalent to the annual electricity needs of more than 2,000 people. These breakthroughs show how offshore wind is evolving to meet extreme weather challenges while delivering abundant, clean power from the sea.[1]

Nuclear energy, which is generated through controlled nuclear fission, remains one of the most reliable sources of low carbon power. In recent years, however, debates about its safety have resurfaced, especially after the Fukushima disaster in Japan, which was triggered by a massive tsunami. It reignited long-standing concerns and brought back memories of the Chernobyl meltdown, which lives on not only as a tragic chapter in nuclear history but also through the television series that you may have watched or had recommended to you by that one friend who devours every disaster-themed show. An encouraging development is that recent innovations have focused on compact and inherently safer reactor designs known as Small Modular Reactors, or SMRs. These reactors can be manufactured in factories and transported to their final location, reducing construction timelines and improving oversight. Many experts also believe that SMRs may allow nuclear power to become more flexible and more distributed, making it possible to bring clean baseload energy to regions where large conventional plants would be impractical. Innovation in nuclear power has accelerated, with thorium-fuelled molten salt reactors (MSRs) emerging as a potentially safer and cleaner alternative. They are resistant to meltdowns and produce significantly less radioactive waste. Since Thorium is also more abundant and carries a lower risk for weaponization, it makes MSR a more secure option too.

Geothermal energy comes from the natural heat deep within the Earth. The deeper you go, the hotter it gets, especially near the Earth's core. In geothermal systems, we drill deep underground to reach hot rocks and underground water. The heat from these rocks warms the water, turning it into steam. This steam comes back up through pipes and is used to spin turbines, which then produce electricity. A newer method called CO_2-Plume Geothermal (CPG) uses CO_2 instead of water. The CO_2 heats up underground and rises to the surface, driving turbines to generate electricity. At the same time, a good amount of the CO_2 stays trapped below, offering the added benefit of carbon storage. Another approach, called pumped geothermal, involves injecting water into hot, dry rocks to create steam in areas without natural underground water.

Methane is nearly 80 times more harmful than CO_2 when it comes to trapping heat in the atmosphere, and is produced by sources like livestock and landfills. Today, innovative bio-reactors, which are essentially small systems that use natural enzymes and microbes, along with catalytic technologies, are being developed to convert methane into water and CO_2. But wait, isn't CO_2 harmful? Less harmful than methane, sure, but what do we do with it?

FRONTIERS OF CARBON MANAGEMENT

Well, let us just say we store this carbon away, either deep underground, or convert it into something stable, like a physical object. There are several technologies today that work to capture carbon directly and store it. For instance, Direct Air Capture (DAC) pulls CO_2 straight from the atmosphere using filters and fans. Advancements in technologies mean that this technology is also being extended to capture carbon from industrial processes. A number of advanced and novel materials are being developed that can help in capturing this carbon (think of these as the filters that let air pass, but stop the CO_2). Such carbon can then be stored underground, including in depleted oil and gas reserves. It could also be converted into building materials such as carbon black (which is used in inks and tyres) or into carbon fibres, which also find use in the aerospace industry.

Today, companies and startups are pioneering unique technologies to capture and store carbon in ways that would have seemed impossible a decade ago. CarbonCure, a Canadian company, injects CO_2 into concrete during the mixing process. This not only reduces the need for cement, which is one of the most carbon-intensive materials used by society, but also locks away CO_2 permanently in a stable and usable form. If you enjoy hiking, you may have noticed that rocks weather over time. This natural process happens when minerals react with rainwater and CO_2, gradually turning CO_2 into solid carbonates. It is one of nature's oldest carbon removal mechanisms, but it works very slowly. Today, many innovators are building technologies to speed it up.

Novel systems such as PeroCycle focus on the steel industry, which contributes close to 8 to 9 per cent of global emissions. Their approach accelerates the reaction between CO_2 emitted by steel plants and specially designed mineral or metal oxide surfaces. This allows the CO_2 to be captured on site and converted into useful materials, turning what was once waste into a resource. Researchers are also exploring how the oceans can help. Some initiatives study whether adding magnesium-rich minerals to seawater can safely enhance its natural ability to absorb CO_2, while keeping ecosystems intact. These efforts are at early stages, but demonstrate how science, engineering and nature-based inspiration are converging to create new possibilities for removing carbon from the air.

GREEN MOLECULES

Green hydrogen refers to hydrogen produced by using clean electricity to split water into hydrogen and oxygen. Because no fossil fuels are involved in the process, it is produced without carbon emissions at the point of generation. This makes green hydrogen a versatile clean fuel for parts of the economy where direct electrification is difficult, including steel and cement production, fertilizer manufacturing, shipping, heavy transport, and as a means of storing surplus renewable energy for later use.

However, hydrogen is a very light gas, which makes transporting and storing it safely over long distances complex and expensive. To solve this, new technologies are being developed to make hydrogen easier to move around the world. One promising solution is the use of Liquid Organic Hydrogen Carriers or LOHCs, which temporarily bind hydrogen to a liquid that can be transported safely in tanks and pipelines. Once it reaches its destination, the hydrogen is released and the carrier liquid can be reused.

Other renewable fuels are emerging that are not only easier to handle but also extremely versatile. Two of the most important among them are green ammonia and green methanol. Both are created using green hydrogen: ammonia by combining hydrogen with nitrogen, and methanol by combining hydrogen with captured CO_2.

Green ammonia plays a central role in global food systems because ammonia is a key ingredient in fertilizers. Producing it without carbon emissions helps make farming more sustainable and strengthens food security. Beyond agriculture, ammonia is gaining momentum as a clean fuel for shipping and power generation. Another advantage is that ammonia can serve as a carrier for hydrogen. Because ammonia can be stored and transported as a liquid, it can be converted back into hydrogen when needed.

This opens the possibility of global trade in green hydrogen, with ports becoming major hubs. In many countries, fertilizer and steel plants are already located close to ports, which means green ammonia can be transported directly to these industries or used as a fuel for ships that dock there. It can also be shipped to countries where cement, steel and fertilizer plants are located inland, helping decarbonize heavy industry far from coastlines.

Green methanol is another versatile fuel. It is used in shipping, road transport and the chemical industry. Because it is made from green hydrogen and captured CO_2, it reduces emissions and allows carbon to be recycled rather than released into the atmosphere. It is also an important building block for plastics and many everyday materials.

Together, green hydrogen, green ammonia and green methanol offer powerful tools for decarbonizing some of the hardest sectors to clean, including heavy industry, long distance transport, shipping, agriculture and chemicals. They create an entirely new energy ecosystem that can operate across borders and continents.

Biofuels form another important pillar of this cleaner energy system. When you sip a glass of whisky, you are tasting ethanol, the same type of alcohol that, in a different form, can power vehicles. Ethanol has long been used as a renewable fuel, especially when made from crops like sugarcane or corn. This is known as first generation, or 1G, ethanol. Today, however, the world is also exploring second generation, or 2G, ethanol which uses agricultural waste such as straw and husk instead of food crops. This approach avoids competition with food production and makes use of vast quantities of biomass that might otherwise be burned in fields. New biochemical processes and advanced enzymes are making this possible at scale. Companies like Chempolis, for instance, have developed technologies that can break down tough agricultural residues into fermentable sugars, enabling cleaner ethanol production with much lower emissions.

Ethanol is also becoming an essential ingredient in Sustainable Aviation Fuels (SAF). These fuels are crucial for decarbonizing aviation, which remains one of the most difficult sectors to decarbonize. Several countries are now introducing mandates requiring airlines to use a minimum share of SAF in their fuel mix over the coming years. These mandates are expected to grow steadily, pushing innovation and investment in advanced biofuels, green hydrogen-based fuels, and other sustainable pathways for aviation.

FRONTIERS OF ENERGY STORAGE

Solar and wind energy do not produce power all the time because the sun shines during the day but not at night, and the wind does not always blow. This means the energy from these sources comes and goes. If we fill the gaps by using fossil fuels like coal, it does not help reduce emissions. What we really need are large storage systems that can save extra energy when there is plenty, and supply it when days are dark and winds are still. This way, we can have clean power all the time.

Think of batteries—giant ones—capable of storing massive amounts of energy that can power towns and cities. They will store renewable power during peak production and discharge it when generation is low, ensuring steady clean energy supply. Lithium-ion batteries are currently the dominant technology for short to medium-duration energy storage, offering high energy density, rapid response times, scalability and rapidly declining costs. This has led to their widespread deployment alongside solar and wind farms worldwide. Several advanced battery technologies are emerging to address various grid storage needs: solid-state batteries provide safer, longer-lasting performance with higher energy density; sodium-ion batteries offer a more abundant and cost-effective alternative to lithium, making them promising for large-scale and stationary energy-storage applications where cost, safety and material availability matter more than energy density; zinc-based and aluminum-air batteries use eco-friendly, sustainable materials aimed at reducing environmental impact; while flow batteries store energy in large external tanks of liquid electrolytes, enabling scalable, long-duration discharge suitable for utility and industrial use.

Now consider the portable battery bank that you use to charge your phone. Can you really expect it to provide you power backup for extremely long durations? Do you think you could work the entire night on your laptop without charging it? That is the challenge with batteries. But what if we store this energy differently?

Today, one of the most impactful innovations in renewable energy storage is pumped storage hydropower. When electricity demand is low and renewables generate excess power, that surplus is used to pump water uphill into a storage reservoir. Later, during periods of high demand, the stored water is released downhill, turning turbines to generate electricity. In effect, this functions like a natural battery. Instead of storing electricity directly, we store potential energy through gravity and water. This is often described as a form of gravity-based energy storage known broadly as a Gravity Energy Storage Systems or GESS.

A powerful example of this is Greenko's integrated renewable energy project in Pinnapuram, India.[2] The facility combines large-scale solar and wind generation with a massive pumped-storage plant that can store several gigawatt hours of clean energy. Since it is a closed loop system, meaning the reservoirs reuse the same water rather than drawing continuously from a river, it avoids disturbing natural river flows and reduces ecological impact.

During the day, excess solar and wind energy pumps water to the upper reservoir. When the grid needs more power, the water flows back down to generate electricity, ensuring round-the-clock clean energy.

Greenko also realized that it could offer this storage capacity as a service. Other renewable energy producers can now use the Pinnapuram facility as a kind of giant battery to firm up their own wind and solar supply, enabling them to offer reliable, round-the-clock clean power without building storage of their own. This creates an entirely new model where long-duration storage becomes a shared grid asset, strengthening the energy ecosystem for everyone.

But once the wheels of innovation are set in motion, there is no going back. Energy Vault, a Swiss company, pioneered an AI-powered technology to lift heavy concrete blocks up to elevated heights.[3] When energy demand rises, the blocks are lowered, converting gravitational potential energy into electricity. This system operates like pumped hydro, storing energy by lifting weights using surplus renewable power and generating electricity as they descend.

Smart grid integration, coupled with AI-driven load forecasting, allows these storage assets to balance supply and demand dynamically, making renewable energy more reliable and flexible across sectors.

FRONTIERS OF UNIFIED ENERGY

Technology created many of the problems we face today but it is also creating the solutions. From steam engines and coal plants to green hydrogen, solar panels and advanced rock weathering, innovation has steadily shifted us towards cleaner alternatives. Now AI is helping speed up everything from designing better batteries to discovering new ways to capture carbon. What once caused emissions may now help erase them. The future may still be powered by the very force that once drove the past.

Clean energy solutions function best when integrated into a cohesive ecosystem. Solar power can generate electricity that can power homes, but the same energy can be used to split water using electrolysis to produce green hydrogen. Green hydrogen, which can be stored and used in petroleum refineries, production of fertilizers, steel and in powering vehicles, can also be used as input for fuel cells which can then produce electricity. One can only think about the various use cases that these technologies could have.

For example, during the day, solar energy could be used to generate and store green hydrogen. Then, at night, when electricity demand rises, this stored hydrogen could be used to power fuel cells, which would convert it into electricity to meet the load.

The transition to net-zero can never be driven by a singular solution, or multiple solutions working in isolation. It requires a system of systems that must also be efficient, inclusive, cost-effective and sustainable. In an ideal world, power generation, storage, carbon capture, sustainable fuels, and built environment technologies must work in unison to address a threat multiplier like climate change. And just as important, the processes involved in making these technologies—from sourcing raw materials to manufacturing and transport—must also be clean, because solving one problem cannot come at the cost of creating another, especially when it worsens the problem that it is itself trying to solve.

But is there something that cuts across all of these sectors and use-cases, something that connects energy, industry, mobility and daily life, and has the potential to act as a true force multiplier? Something that, when combined with the many technologies we already have to fight climate change, can help the developed world reduce its carbon footprint, and allow the developing world to grow without following the same high-emission path?

There is. It is transforming the world as we know it today. And chances are, you use it every single day. *AI.*

PART 2
AI × CLIMATE

8

THE AI REVOLUTION

Powering Humanity's Greatest Leap

The night before the fire, Sophie was working on an assignment. It had been given to her class two months ago, but just like most of us during our school and university days, she finally started working on it the night before. She turned on her laptop, typed the question into ChatGPT, and asked it to generate an essay. Within seconds, she had one. She spent a few minutes cleaning it up and making some edits, but within 30 minutes, it was complete.

On a side note, if you are a student reading this, you might be surprised to know that those of us who went to school and university before ChatGPT also managed to submit assignments the night before. It simply took more time, more stress and significantly more caffeine.

Later that evening, she went downstairs to join her parents at the dinner table. Her mother had just ordered some new furniture and was telling her father how her Instagram feed had been showing her the most beautiful rustic-modern pieces lately. She was impressed by how stylish they looked and surprised that so many people seemed to be interested in the same aesthetic. Sophie explained that this was not a coincidence. Social media platforms use artificial intelligence to personalize the content each user sees. Because her mother had searched for that kind of furniture, even once, the system detected her interest and started showing her more of the same. These algorithms also track how long a person watches a video before scrolling away. So, if she paused on a reel featuring rustic decor, the platform would take note and begin prioritizing similar content in her future feed. Sophie added that while her mother might think everyone else is seeing these posts too, this personalized content is actually unique to her feed. AI tailors what each person sees based on their individual browsing habits and interactions, so not everyone views the same content. Her parents were both surprised by this.

As the conversation switched to AI, her father shared that his company had recently adopted AI-enabled software that identifies fraudulent insurance claims with impressive speed and accuracy. He expressed surprise at the system's success rate and explained how much money the company had saved by automatically rejecting false claims that would have previously slipped through. However, he lamented that some of his colleagues, who had been responsible for manually reviewing claims, had been let go as a result of this transition. The company was also hiring new professionals whose job was to develop strategies for mainstreaming AI across other departments. While some roles had been lost, new ones were being created.

After dinner, Sophie returned to her room, gathered some laundry, and tossed it into the washing machine. A sticker on the front proudly declared that the machine uses AI to detect stains and automatically adjust water levels and temperature. She cracked open the small basement window, hoping to let in a bit of cool air, though the heat lingered heavily. If she had paused for a moment and looked outside, she might have noticed a faint, eerie red shimmer along the distant sky. Miles away, a wildfire had begun its slow advance, its flames feeding on parched vegetation after an unbearably hot summer and propelled by restless winds. Little would she know that within a few hours it would sweep through and engulf her entire neighbourhood.

AI EVERYWHERE

The scale of AI's presence and influence is difficult to overstate. As AI continues to embed itself across every sector, from education and healthcare to consulting, climate action and governance, there will come a time when the term 'AI' will no longer serve as a distinguishing feature. It will simply be the default. Everything will, in some form, be powered by artificial intelligence.

In healthcare, AI is significantly advancing medical diagnosis and patient care by rapidly analysing complex data more accurately than many traditional methods. For example, AI tools assist radiologists in detecting breast cancer from mammograms with improved precision and faster diagnosis than manual readings alone. In intensive care units, AI-driven predictive models monitor real-time patient data, combining vitals with lab results, enabling real-time interventions, which would otherwise have taken precious hours. Wearable devices equipped with AI continuously track heart rhythms and alert users to irregularities, sometimes automatically notifying emergency services during critical events like cardiac arrest, facilitating rapid medical response. Similarly, AI-powered systems help detect diabetic retinopathy in eye scans earlier, reducing the risk of blindness. One of the most groundbreaking AI achievements is Google DeepMind's AlphaFold, which solved the protein folding problem–predicting the three-dimensional structures of proteins from amino acid sequences.[1] This breakthrough, recognized by the 2023 Nobel Prize in Chemistry, accelerates drug discovery, deepens disease understanding, and opens new possibilities for targeted therapies, transforming biomedical research. While AI does not replace medical professionals (though we can't really be absolutely sure about anything when it comes to AI), it is becoming increasingly indispensable as a tool that enhances clinical decision-making, enables earlier and more precise diagnoses, supports personalized treatment plans, and accelerates research efforts.

AI-powered tools can screen resumes, assess facial expressions in interviews, and generate tests tailored to candidates. But then, as is the case with technology, people have begun cheating AI-powered video interviews by feeding model answers into tools that mimic eye contact and vocal tone. AI is both the gatekeeper and, increasingly, the force we must carefully manage. Even as universities across the world witness a surge in AI-assisted submissions of student assignments and seek ways to tackle it, perhaps returning to pen and paper for assessments may not be such a bad idea after all.

Today, chatbots can understand context, language, tone and intent, delivering instant resolutions 24/7. Many customers can no longer tell whether they have interacted with a human or a machine. Interestingly, while these chatbots are widely deployed by call centres and the private sector, their use by the government is increasingly gaining traction. Many government departments spend countless hours managing citizen complaints. AI can relieve them of this grunt work, freeing staff to be re-skilled or up-skilled and spend more time on field visits and strategic planning. This will allow human beings to do what we do best—connect with each other, and form closer links to the communities we serve.

AI models are now capable of predicting market trends and detecting financial fraud with remarkable precision. Beyond these advanced functions, AI is also becoming a powerful tool to promote financial inclusion and support micro-entrepreneurship. One of the primary barriers that prevents low-income individuals from starting their own businesses is the issue of credit-worthiness. In the absence of collateral, banks are often hesitant to issue loans. AI-powered tools can help address this challenge by analysing spending behaviour, transaction histories and other financial patterns to generate alternative credit assessments. This approach provides banks with a more comprehensive understanding of a potential borrower's financial responsibility. While such methods may raise questions around ethics and transparency, they represent a significant improvement over the traditional practice of automatic rejection based solely on the lack of collateral. With ongoing refinement, these systems could significantly increase the chances of underserved individuals receiving access to credit. In doing so, they would help unlock opportunities for small business creation and contribute meaningfully to economic empowerment.

In agriculture, AI-enabled drones monitor crop health from above, while simple photographs taken by farmers can be uploaded and analysed to detect diseases, assess plant vitality and recommend targeted interventions.

On the climate front, AI is playing a critical role in how we understand and respond to environmental and urban challenges. It is being used to model climate risks with greater accuracy, helping scientists and governments predict extreme weather patterns, including the likelihood and severity of floods. AI can optimize energy consumption by managing electricity grids, adjusting lighting and cooling based on demand, and reducing overall waste. AI tools guide emergency response teams by mapping risk zones, improving evacuation planning and allocating resources in real time. Urban planners increasingly rely on AI to simulate traffic flows, redesign transport systems, and reduce emissions through intelligent routing. AI is also used to monitor air quality, detect pollution hotspots and suggest targeted policy interventions. On the whole, AI can make our cities and their governance a lot smarter, and a lot more efficient. Even data centres, which form the backbone of the digital economy, are now being managed more efficiently with AI helping them maintain performance and reduce energy strain during periods of extreme heat. AI is no longer simply a support technology. It is becoming a central force in building resilient, adaptive and sustainable cities.

The range of applications is vast and constantly expanding. In fact, one could easily dedicate an entire volume to AI use-cases alone. Yet, as AI becomes embedded across every sector and technology, the need to catalog its applications may eventually diminish. What we now describe as 'AI-powered' may soon become the default, while technologies that do not incorporate AI may quietly become obsolete. The distinction between AI and non-AI solutions is beginning to blur. Over time, it may disappear altogether.

Think of AI as a universal force, operating horizontally across diverse sectors through generalist chatbots, and vertically by penetrating deeply into specific functions such as monitoring crop health. This pervasive reach makes one thing clear: there is no turning back.

NO GOING BACK

AI is slowly becoming as invisible and as indispensable as electricity. We cannot see electricity, but can we imagine a world without it? It silently powers our lives, from public transport to refrigerators in our kitchen, from keeping our phones alive to keeping the internet running, and from helping run air conditioners in an ever-warming world to keeping stadiums lit during matches. As electricity reached previously unconnected communities, it unlocked new waves of productivity. Cold storage extended the life of produce so markets could grow, power looms multiplied textile output and shifted labour into factories, and electrified roads, factories and communication networks enabled longer hours, faster logistics and entirely new forms of coordination. Productivity rose not simply because machines ran, but because people, places and systems could finally connect and operate at a scale that earlier generations could not even imagine.

AI today is beginning to play a similar role. If electricity was the great enabler of the twentieth century, AI is becoming the great enabler of the twenty first. It is turning into a kind of intelligent grid for the world, quietly embedding itself into decisions, systems, services and tools that shape how everything functions. And just as electricity required fuel to generate power, the AI ecosystem is fuelled by GPUs that act as the engines of computation. If the electricity grid could speak, it might not be entirely pleased with this comparison, especially considering how much power AI now pulls from it, but that is a story for a later part of this book.

According to McKinsey,[2] 78 per cent of organizations reported using AI in 2024, up from 55 per cent just two years earlier. In response to a survey by McKinsey,[3] 65 per cent of the organizations also revealed that they were now regularly using generative AI, a figure that stood at about 33 per cent in 2023. As expected, many of the respondents also revealed that AI had reduced costs, and increased revenues. A back-of-the-envelope estimate may show that based on results from various surveys, almost 2 billion people across the world have used AI, with about 500 million people engaging with it on a regular basis—a number that is higher than the population of the EU. Understandably, this pace of change has triggered concern. Ethics, data misuse, misinformation, job displacement and the energy footprint of large-scale models are all real issues associated with AI. If one were to undertake a survey in any random subset of people, these are all likely to be flagged by respondents as key issues.

We have all read history, and if we still remember some of what we learned at school, the printing press was once feared for disrupting religious authority, but ended up bringing about an information revolution. Some thought electricity was too dangerous. The internet was met with concerns about fraud, privacy, surveillance and pornography, and concerns continue to emerge. But this is the nature of technology. Every major leap in technology brings uncertainty, yet, none of these technologies were reversed. Instead, societies adapted, developed safeguards and moved forward.

Skepticism about technology is natural, but history has repeatedly shown that technology often becomes the solution to the very problems it creates. It was industrial technology that polluted the skies with coal, but it is now green hydrogen, solar energy and carbon capture systems, each a product of advanced engineering, that are leading the transition to cleaner energy. The same is true in the digital realm. AI can generate deepfakes and misinformation, yet it is also used to detect manipulated images and flag false narratives at scale. AI can be exploited to breach cybersecurity systems, but it is also deployed to detect anomalies and stop cybersecurity threats before they escalate. Yes, there are valid concerns around ethics, privacy and control. But regulations are already evolving, and so too is public awareness. The answer is not retreat. It is to find the sweet spot where innovation is balanced with responsibility, and where technology is guided by human intent rather than fear.

The determining aspect about technology are the externalities, or the spillover effects. Do they only impact what they were developed for, or do they have a wider impact? Clearly, it is the latter. The Industrial Revolutions did not only change how goods were produced. They changed how people lived. Steam engines revolutionized travel. Electrification brought light to homes and extended the day. The internet reshaped communication, education and the economy. Along similar lines, AI is already beginning to do the same: redefining how we work, learn, govern and solve complex problems. Therefore, we need to understand that the disruption brought about by AI is not just a technological shift. It is a societal one. Countries that embraced previous revolutions did not look back. They used the moment to invest, reform and reimagine their futures. AI now presents a similar moment. It is not merely a tool. It is the foundation of what many are calling the Fifth Industrial Revolution.

The question is no longer whether AI will shape the future. It is how we choose to shape AI, and that includes embracing both challenges and opportunities, and converting even the challenges into opportunities.

THE GLOBAL OPPORTUNITY

Efforts must be made to shape AI positively, inclusively, ethically while ensuring that it is for all, and that it is used to not only drive economic growth, but also accelerate digital inclusion and socioeconomic development. Despite all concerns (and as we have flagged, technology will also give solutions to the challenges that it brings), there is a reason to be bullish.

According to PwC, AI could contribute up to USD 15.7 trillion to the global economy by 2030, equivalent to about 14 per cent of global GDP and one of the largest commercial opportunities in human history.[4] McKinsey's analysis shows that generative AI, together with other advanced learning systems, could raise global productivity by 0.5 to 3.4 percentage points each year, unlocking significant economic value across sectors.[5] Massive disruptions are already underway. The global race continues to get heated up. Around the world, companies, countries and compute giants are pouring unprecedented sums into building the next generation of AI infrastructure. In India, Google recently announced one of its largest commitments anywhere in the world: an AI city anchored by a 1 gigawatt data centre in Visakhapatnam, backed by an investment of roughly USD 15 billion over the next five years.[6] This was followed by Microsoft announcing a USD 17.5 billion investment in India to drive AI diffusion at population scale.[7] This investment is designed to support advanced AI development, hyperscale computing, and the rising digital economy of one of the fastest growing technology markets on the planet.

Even traditional engineering and infrastructure firms are beginning to reposition for the AI shift. Larsen & Toubro (L&T), one of India's largest engineering and construction conglomerates with deep capabilities across power, industry, and urban infrastructure has signalled a strategic push into digital infrastructure alongside its core operations. The move is notable for what it represents. Large industrial players in emerging markets are increasingly viewing data centres as part of a broader infrastructure stack that spans power, cooling, construction, and long-term operations. In that sense, L&T's entry reflects how data centres are no longer seen as niche digital assets, but as a new class of core infrastructure attracting players far beyond the technology sector.[8]

At the same time, OpenAI and NVIDIA have unveiled an agreement that has captured global attention. NVIDIA has signalled its intent to invest up to one hundred billion dollars in OpenAI as both companies work together to deploy at least ten gigawatts of AI data centre capacity. This scale implies the installation of millions of high-performance GPUs over the coming years.

Google is now likely preparing to take its home-grown Tensor Processing Unit (TPU) to market, signalling an ambition to make its internal AI acceleration stack available to external customers. For more than a decade, TPUs have powered core Google services such as Search, Photos, Maps and the latest Gemini models, functioning as Google's own equivalent of the GPUs that drive modern AI. As reports of a possible commercial rollout surfaced, investors began reassessing the balance of power in the AI chip landscape, with heightened attention on how a mature TPU platform could reshape competition with Nvidia.[9] These announcements mark one of the largest waves of compute and energy infrastructure investment in history. They underline a simple reality: the future of AI is no longer shaped only by algorithms, but by those who can build and power the immense physical systems that AI requires.

As per the LinkedIn Economic Graph,[10] a real-time, digital map of the global labour market, built by LinkedIn using the massive volume of data it collects from its users, job postings, companies and educational institutions, 70 per cent of job skill requirements are expected to evolve due to AI by 2030. The International Monetary Fund has warned that 40 per cent of global jobs will be affected by AI, creating both disruption and opportunity. At the regional level, countries that invest in AI could see GDP increases of up to 26 per cent, according to long-term digital transformation forecasts. This signals a deep structural shift. AI is rapidly becoming a foundational layer of the global economy, much like electricity, the internet or mobile technology once became. For countries, businesses and individuals, the question is no longer whether to adopt AI, but how to harness its full potential. Those who move early and decisively are likely to lead in productivity, innovation and competitiveness. *Those who hesitate risk falling behind.*

An uncertain world order challenged by multiple crises may seem bleak, but history shows that adversity often sparks transformative change. Countries, businesses and individuals have repeatedly turned challenges into growth opportunities. After the Asian Financial Crisis in the late 1990s, South Korea rebounded by investing heavily in innovation, education and technology-led exports. Within two decades, it became a high-income, high-tech economy and is now a global leader in electronics, automotive manufacturing and digital infrastructure. During the COVID-19 pandemic, streaming platforms rapidly expanded by increasing content and localizing offerings, reshaping how people consume media worldwide. Remote work tools such as Zoom and Microsoft Teams transformed the workplace, while educational technology reached millions and accelerated digital transformation at a pace few anticipated. Similarly, the European Union's response to the 1970s oil shocks led to major long-term reforms in energy policy and efficiency standards. What began as crisis management eventually drove sustained investment in renewable energy, energy diversification and technology research, which today continue to influence Europe's climate and energy leadership.

AI FOR CLIMATE?

At the time of writing this, Texas is facing flooding which scientists describe as a once in a millennium event, intensified by hotter and more humid conditions linked to a changing climate.[11] In Europe, a recent heatwave claimed over 2,300 lives, with attribution studies showing that climate change made much of this tragedy far more severe.[12] In Brazil, record-breaking rainfall in the southern states displaced hundreds of thousands of people and pushed local infrastructure to its limits, with researchers finding that human-caused warming had sharply increased the likelihood of such extremes.[13] These events, occurring within months of one another across different continents, offer a stark reminder that climate-driven disasters are no longer rare anomalies. They are becoming part of the lived reality of our time, underscoring the urgent need to strengthen global resilience and response.

AI has become a powerful force today, offering immense potential. But can it be harnessed not only for innovation and growth but also to combat climate change? The answer is complex. Training large AI models consumes vast energy and water, and AI data centres can be carbon-intensive. Though efforts to decarbonize these facilities are underway, including innovative approaches covered in later chapters, the rising AI demand is leading to more data centres cropping up, sometimes in areas facing water shortages and sometimes in areas where the electricity grid barely has any renewable energy.

Conversely, AI accelerates climate solutions by enhancing climate models, detecting illegal deforestation, optimizing energy grids, designing advanced solar materials, forecasting extreme weather, improving irrigation and simulating carbon capture. In this way, technology itself aids the fight against climate disruption. The clear opportunity is that AI, if managed wisely, can be a vital tool in addressing this escalating crisis.

In the chapters ahead, we will take you on a fascinating journey through the two sides of AI's impact on climate change. First, you will discover incredible ways AI is powering breakthroughs and fuelling real progress in our fight to protect the planet. Then, we will explore some of the challenges and risks that come with this powerful technology, particularly on the climate and sovereignty front.

But don't worry. This is no doom-and-gloom story. Ultimately, we get into solutioning, where we share practical, inspiring ideas for how we can work together to leverage the AI–Climate nexus to unlock huge benefits for society and the environment.

In the next chapter, we explore how this extraordinary technology can become one of our greatest allies in tackling climate change.

9

PREDICTING THE UNPREDICTABLE

AI for Climate Resilience and Adaptation

Thousands of miles from Sophie, and just as far from Asha, cradled deep within a landlocked nation, sits the small and unassuming city of Climaville. Its residents enjoy a peaceful life, although the distant thunder rolling from two different parts of the world—Asha's and Sophie's—has not gone completely unnoticed here. Droughts and conflicts are scorching the land on one side while wildfire rages on in the other, both triggered by climate change. One side calls climate change a hoax while waging trade wars, the other struggles to dislodge militias blocking a vital transcontinental trade route, disrupting major supply chains. The country where Climaville lies has so far remained relatively untouched by the crises unfolding elsewhere.

But not every crisis stops at a border, least of all climate change. Today, there is thunder-rolling in Climaville, real thunder. It has been raining all night, and it is turning into a storm, fast.

Ori woke to the sharp buzz of his alarm, still heavy-eyed from a late night. A final year university student, he carried a deep concern for climate change and, like many of his generation, a growing fascination with AI. The night before, he had been absorbed in a project that used natural language processing to sift through social media posts and news reports, searching for early signs of deforestation and rising river levels, patterns that, if found in time, could give communities the precious warning they needed to prepare.

He got ready for university, grabbed his bag, slipped on his rain jacket, and stepped out into the rain, which had slowed to a drizzle now. His thirty-minute walk to campus took him through a nature trail, and past the river. On most days, he would pass stalls selling fresh farm eggs and enjoyed cheerful exchanges with the farmers. But no one had set up shop today because of the incessant rains. Climaville was a beautiful city, surrounded by hills on three sides. As he crossed the old iron bridge, car drivers waved in greeting; everyone knew everyone in their town. But as he looked down to the river, he noticed how high the current had risen overnight. In the town's library, Ori had once read about floods from centuries ago, but there had never been one since. This felt strange and unsettling. The water was well above the painted danger line, a level he had never seen breached before.

Only a few minutes after he stepped off the bridge, a deep rumbling rolled through the valley. He watched in disbelief as torrents of water gushed down the hills and through the valley. The bridge gave way, taking with it a car whose occupant had no time to escape. The flash floods tore through the streets, roads disintegrated, power snapped and even the newly built data centre, the city's pride—opened with much fanfare only months ago—was flooded. The cellular signal vanished. When emergency crews did arrive, they struggled to coordinate without power and network.

The mayor was in shock, unable to fathom the scale of what Climaville had experienced. What they had witnessed was a phenomenon known as cloudburst, which is when massive rain falls in one spot—all at once—almost like the sky has upended an entire jug of water in a single pour. It happens when warm, moist air shoots upward, cools quickly, and dumps all its moisture in minutes instead of hours. Scientists warn that these sudden cloudbursts are likely to happen more often with climate change.

But what if this could have been predicted? Climaville was not ready for this calamity, but could it have been?

Had the data of rainfall history, soil saturation, topographic risk, real-time atmospheric conditions, satellite imagery and structural integrity assessments of local infrastructure been synthesized and interpreted by advanced AI systems, the story could have ended differently. AI has the capacity to process thousands of data layers and uncover weak links no human analyst could in a very short timeframe. It can identify risks, model cascading failures and simulate scenarios that could give cities a vital headstart. In the case of Climaville, it certainly could not have stopped the rain. But it could have warned the mayor and the residents that their town was now at an elevated risk of phenomena such as a cloudburst, and that it, in turn, could lead to flash floods. It could have warned about the cracks developing in the bridge, and it could have predicted which routes the water from flash-flooding might take, preventing the construction of electrical substations and data centres along those areas. It could have helped emergency services know where to position resources, and help communities get out of harm's way before the waters rose.

Climaville is fictional, but its story is not. Cities around the world are already living through variations of this exact sequence. From India to the United States, from Brazil to Italy, once-in-a-century disasters are becoming frighteningly regular.

In earlier chapters, we examined the far-reaching impacts of climate change across geographies, across sectors, and through their many interconnected dimensions.

In the previous chapter, we explored how AI is transforming the world and the potential it holds for tackling climate challenges. Now, we dive into how AI can be used to build climate resilience and drive climate adaptation efforts.

As a sidenote, we did not choose the name 'Climaville' so we could depict it as a victim of climate change. We chose it because, by the end of the next few chapters, we want it to represent what a city leading the fight against climate change, using AI, could look like.

UNDERSTANDING RESILIENCE AND ADAPTATION

Resilience refers to the ability to withstand and absorb shocks. In the case of climate resilience, the meaning does not change. It essentially refers to the capacity of infrastructure, communities, ecosystems and economies to prepare for, withstand and recover from climate-related shocks and stresses. Floods, heatwaves, cyclones could all be examples of such shocks. Adaptation means the capacity to adjust to, or cope with the changes that these stresses and shocks bring about. In the city of Climaville, resilience would have meant designing bridges that could withstand flooding of a certain intensity and having a backup communication system that could kick-in if the primary system failed. If flooding became a frequent phenomenon in Climaville, adaptation would have meant creating elevated walkways that would allow citizens to cross roads that get flooded frequently and setting up alarm systems that can warn citizens to reach higher ground ahead of flooding. Adaptation, in a way, is a part of resilience efforts.

In Sophie's part of the world, a region increasingly scarred by wildfires, the signs of a changing climate were already part of daily life. This summer was abnormally hot, and resilience and adaptation could have entailed designing houses that are cooler. Perhaps they could have painted rooftops to reflect sunlight or constructed houses with fly ash bricks, which keep the indoor temperature cooler. In Asha's drought-stricken part of the world, it would entail the use of locally suited drought-resistant crops.

Today, AI is helping us see what lies ahead. By analysing layers of complex data across a number of fronts—environment, geology, infrastructure, weather—AI offers an unprecedented ability to anticipate, adapt and act.

Here are some of the most transformative ways AI is being used to build climate resilience:

AI FOR EXTREME WEATHER FORECASTING

Advanced AI models can power early warning systems that detect cyclones, heatwaves and floods, hours or even days before they strike. These models draw on satellite data, historical records and real-time weather feeds to offer alerts with remarkable precision. In South Asia, for instance, Google's Flood Forecasting System has helped millions in India and Bangladesh receive accurate, hyperlocal flood warnings, saving lives in some of the most flood-prone regions in the world.[1] As Storm Bebinca approached Taiwan in 2024, Taiwan's Central Weather Administration (CWA) used AI to predict the cyclone's path with a high degree of accuracy. Interestingly, Typhoon Gamei, one of the strongest to strike Taiwan, was also tracked using AI-based weather models. It was able to predict a hit almost 8 days before the cyclone made a landfall.[2] Such early warnings play a vital role not only in emergency preparedness and response, but also in safeguarding lives and critical infrastructure. With an advance notice, authorities can activate evacuation plans, secure essential services, and even protect supply chains by rerouting logistics or relocating goods away from high-risk zones. By analysing historical storm data and real-time atmospheric patterns, AI capability can help with timely alerts, better-targeted evacuations and reduction of economic losses through appropriate interventions in the supply chain.

AI FOR BUILDING AND MAINTAINING DISASTER RESILIENT INFRASTRUCTURE

Today, we can combine various data layers, including Geographic Information Systems (GIS), to organize and analyse information linked to specific locations, much like a city planner using a detailed street map to decide where to build schools or parks. By applying powerful AI algorithms, we can create digital twins of entire cities. Think of a digital twin as a 3D model in front of you. You can use it to simulate scenarios, you can use it to predict what would happen if you move something around, and you can also use it to figure out how to evacuate people during an emergency. These AI-powered digital twins simulate how buildings, bridges and roads would respond to stress caused by extreme heat, floods or seismic activity. These simulations allow engineers and planners to make infrastructure more robust before a disaster hits. Think of digital twins as advanced flight simulators—the ones that you may have seen on TV—except that in this case, they can be created for everything, right from a small warehouse to an entire city, and even a country. In Lisbon, city planners used a digital twin developed by Bentley Systems to map stormwater flows and test infrastructure scenarios, enabling them to pinpoint where to build intercepting tunnels and strengthen the drainage network. The modelling suggests the city could avoid significant flood events over the next century and reduce damage to infrastructure and livelihoods by over €100 million. The project stands as an early example of how digital-twin technologies can shape urban resilience.[3] AI-powered sensors can also detect vibrations, stress changes and micro-fractures in bridges and dams, allowing for preventive maintenance before collapse. These models are particularly useful in flood-prone regions, where infrastructure is often under immense pressure.

AI FOR MANAGING WILDFIRE-RELATED RISKS

We are seeing a growing number of wildfires around the world, particularly during dry seasons. Remember what happened to Sophie's house? These fires have led to the loss of hundreds of lives, the destruction of thousands of homes, and have caused economic damage amounting to billions of dollars. AI systems can analyse wind speed, vegetation dryness and satellite imagery to forecast wildfire outbreaks. By identifying risk zones early, authorities can issue evacuation alerts and build their strategy to respond, with measures such as positioning firefighting resources to prevent widespread damage ahead of these fires reaching major cities. A powerful real-world example comes from California, where SensoRy AI, created by Ryan Honary and adopted by the Orange County Fire Authority, uses a network of ground sensors and machine learning to detect fires when they are smaller than a square foot, providing invaluable lead time for response.[4] Similarly, researchers at the University of Southern California have developed a model that combines generative AI and satellite imagery to predict wildfire spread in real time, helping firefighters stay ahead of flames and improve safety planning.[5]

AI FOR MANAGEMENT OF EXTREME HEAT

Cities, which accommodate millions of citizens—and are likely to witness the inflow of millions more with increasing rural-to-urban migration—are witnessing their infrastructure getting overstressed. These cities are also turning into urban heat islands with the combined effect of construction, heat from air conditioners and vehicular exhaust fumes. The situation is even worse in countries that rely heavily on diesel generators for power. In such cities, AI can help map and monitor urban heat zones by analysing materials, greenery, building density and traffic data. This allows urban planners to design cooler spaces with more trees, reflective surfaces and better airflow. Satellite imagery, analysis of albedo effect (the ability of surfaces to reflect sunlight; the darker the surface, the more heat it will absorb), and other data points such as wind, weather and traffic flows combined can also help in identification of houses that are most likely to bear the brunt of extreme heat conditions. When combined with socioeconomic data, such models can reveal those who are poor, and also the most vulnerable to extreme heat. This is important since poorer households often lack the resources, including access to medical care, to cope with heatwaves. This helps cities identify, prioritize and support the most vulnerable, through measures such as painting rooftops with heat-reflective materials, promoting the use of cooler construction materials, and setting up relief camps with medical assistance in the immediate vicinity of such vulnerable households and neighbourhoods.

AI FOR PROTECTING AGRICULTURE AND FOOD SECURITY MANAGEMENT

In a world where extreme weather events triggered by climate can have serious impacts by affecting agriculture and hitting the food supply (which is also what happened in Asha's village), forecasting crop cycles and deploying remedial measures becomes essential. AI systems can study soil composition, rainfall patterns and genetic crop traits to recommend the best sowing and harvesting windows. This helps farmers protect their yields from erratic weather shifts.

In Cameroon, a mobile application powered by AI is transforming smallholder agriculture. The app allows farmers to upload photos of disease affected crops and receive instant diagnoses along with treatment recommendations. By detecting plant diseases early, the tool helps reduce crop losses and improve yields. It can also function offline making it highly accessible in areas with limited or no internet connectivity.[6] Precision agriculture, powered by AI, is fast and can measure soil moisture, evaporation rates, and also the plant needs to deliver exactly the right amount of water. This is crucial in drought-prone areas where every drop counts. AI tools can predict crop failure due to extreme heat, pests or floods. Combined with other data sources such as those related to floods and cyclones, these models can also predict how and when supply chains are likely to get disrupted, and provide recommendations to optimize supply chains well in advance. This allows governments and humanitarian agencies to plan for shortages before they spiral into crises.

AI FOR MANAGEMENT OF FLOOD RISK

AI models can be used to simulate how floodwaters may move through a city, helping determine where buildings, roads and neighbourhoods should or should not be developed. In the Philippines, the University of the Philippines' Project NOAH (Nationwide Operational Assessment of Hazards) has integrated AI and digital technologies into its flood-hazard mapping strategy. During Super Typhoon Carina, the real-time flood maps attracted over 2.6 million downloads in just two days, helping residents avoid high-risk areas. The use of AI and intelligent modelling has also helped reduce yearly disaster fatalities from approximately 1,200 to 250.[7] This demonstrates how even simple, scalable AI-driven tools can significantly enhance public safety in the face of intensifying climate threats. This could certainly be an interesting tool for Climaville. In Rwanda, an AI-driven flood forecasting model has been built by Microsoft Research Africa in partnership with Carnegie Mellon University. It uses deep learning on meteorological, hydrological, topographic and land-use data in an attempt to predict daily flood extents more accurately, enabling timely alerts and response mechanisms.[8]

AI can also forecast long-term sea level rise by processing satellite imagery, water-level trends, and coastline erosion. This can help guide long-term planning for cities at risk of inundation and help prevent unmanaged migration to low-lying areas. This is especially important as many parts of the world have already started getting affected by rising sea levels, from Venice to Dar es Salaam to Jakarta.

AI FOR CLIMATE RISK MODELLING AND SOCIAL PROTECTION

Imagine if we could know in advance that a dangerous heatwave is coming. AI makes this possible by spotting patterns in weather, temperature and past events to give early warnings. Think how useful that could be: if the government knows people cannot work outdoors for several days, it could send money to daily-wage earners in advance, while hospitals could prepare proactively by ensuring enough doctors, beds and supplies. By predicting these conditions early, AI gives officials the time to prepare smart plans, allocate funds and protect the most vulnerable before the crisis hits.

Insurers and governments can also use AI to understand risk by looking at weather, location, building safety and past disasters. This helps set fair insurance prices and send financial help where it is needed most. For example, One Concern uses artificial intelligence to build detailed digital twins of real-world infrastructure. These virtual replicas, which we described earlier, can simulate how power grids, transport networks and supply chains behave under different stress scenarios. By analysing trillions of data points, the platform helps insurers, utilities and city planners identify vulnerable points long before they fail. It also shows how a disruption in one part of the system can cascade through others, allowing organizations to prepare more effectively and strengthen resilience.[9] Pula, an insurtech company in Africa, uses AI with satellite and field data to offer affordable crop insurance against climate risks like drought. Led by Thomas Njeru, their approach has enabled million in payouts, helping farmers recover faster and encouraging banks to offer more loans. On average, insured farms report up to a 56 per cent increase in yields, boosting income and resilience to future shocks.[10]

AI FOR AIR QUALITY AND POLLUTION MONITORING AND FORECASTING

Across the world, and especially in developing countries, air pollution has reached alarming levels, with 99 per cent of people breathing air that fails to meet WHO standards. It causes about 7 million premature deaths each year, a vast majority of them occurring in low- and middle-income countries.[11] In cities such as Delhi, air quality can reach over twenty-five times the safe limit, and children in polluted areas often have reduced lung function. The toll is silent yet staggering, and most cities in emerging economies lack proper air quality monitoring. AI can help change this. Low cost, decentralized sensors placed on homes, telecom towers, streetlights and moving vehicles can collect thousands of real time data points. Combined with wind patterns, satellite imagery and traffic data, AI can create hyperlocal air quality maps that pinpoint pollution hotspots and generate valuable insights.

For instance, today, it is possible for models to not just predict air quality trends and hotspots well in advance, but also, show, in real time, how to attack them. Such a model could show that if the air quality on a particular day is 200, then of that 200, PM10 is a leading contributor, and 90 per cent of that PM10 is coming from construction in a particular part of a city. Therefore, if we were to intervene there, we would probably witness notable improvement in air quality. This is especially important for developing countries where city municipal agencies suffer from lack of data and lack of funds, and hence, instead of spreading themselves too thin and undertaking piecemeal interventions which do not make a major impact, they could take data-backed targeted actions that would be more impactful.

With targeted action leading to emissions reduction, health outcomes and economic productivity can also be strengthened by protecting the workforce. In India, a Google and Aurassure AI-enabled air quality monitoring system has improved air quality by up to 50 per cent in Chhatrapati Sambhaji Nagar.[12] In South Africa, researchers have developed an AI-enabled system called Ai_r to monitor air quality in real time. The system uses low-cost sensors deployed around Johannesburg, feeding data every five minutes to a cloud-based AI model that predicts pollution hotspots.[13]

AI FOR CLIMATE AND HEALTH SOLUTIONS

AI is emerging as a powerful tool at the intersection of climate and health, helping governments and cities anticipate and respond to climate-linked illnesses. As shifting weather patterns alter the spread of diseases like malaria and dengue, AI models can help with the analysis of real-time data on temperature, rainfall and mosquito-breeding cycles to issue early warnings, which in turn, help in preparedness of public health systems. AI-powered dashboards can also be used to monitor heat stress, air pollution and hospital admissions, enabling more effective allocation of healthcare resources, thereby protecting vulnerable populations. Through accurate predictions made well in advance, resilience of health supply chains can be strengthened while ensuring that medicines, diagnostics and personnel are strategically positioned ahead of climate shocks.

In Vietnam, researchers at the Oxford University Clinical Research Unit (OUCRU) have developed wearables and analytics tools that analyse real-time patient data through AI-powered analysis to detect early signs of severe dengue.[14] As climate change fuels flooding, which increases the incidence of dengue, such tools enable timely clinical responses and improve hospital readiness. Meanwhile, in Benin, a coastal nation in West Africa, the government has adopted an AI-powered malaria diagnostic platform designed for ease of use in low-resource settings. Equipped with automated microscopy and remote diagnostic features, it can be deployed in areas lacking fully developed medical infrastructure.[15]

AI can play a major role in helping health systems prepare for climate risks. For instance, it can guide decisions on where to build hospitals by identifying locations less prone to flooding. It can also give early warnings about heatwaves or cyclones, so hospitals and clinics can get ready in advance. In the case of air pollution, which is a major health problem, AI is already being used to track and manage air quality. These examples show that AI can be used in practical ways across the entire healthcare system. It can help us predict disease outbreaks, support doctors and nurses on the frontlines, make hospitals stronger and safer, and ensure that medical resources are sent where they are needed most.

AI FOR AFFORESTATION AND REFORESTATION PLANNING

Forests play a critical role in preserving nature and keeping climate change in check. They store vast amounts of carbon, clean the air, preserve biodiversity and sustain essential ecosystems. However, the world is losing forests at an alarming pace due to expanding agriculture, mining, and widespread illegal logging. Climate change itself is impacting forests. Wildfires destroy forests and excessive rainfall and flooding events, triggered by climate change, cause landslides that wash away forests along hills and slopes. So, while forests help in combating climate change, in addition to human activities, climate-triggered events are negatively impacting forests, creating yet another *vicious cycle*. Such loss of tree cover is degrading natural landscapes, eroding biodiversity and eliminating some of the planet's most effective natural mechanisms for carbon management.

In response, AI can be leveraged as a powerful tool in afforestation and reforestation planning. AI can identify degraded tracts of land that are most suitable for restoration. It can predict which tree species are most likely to survive under future climate conditions, and it can help monitor the health and growth of newly planted forests. In Brazil, the National Institute for Space Research (INPE) employs AI-enhanced satellite systems to detect and respond to illegal deforestation in the Amazon in near-real time, bolstering protection of reforested zones. According to INPE data, deforestation in the Brazilian Amazon fell by nearly 40 per cent in the first quarter of 2024 compared with the same period the year before. And in parallel, an initiative called re.green uses artificial intelligence, drones and satellite imagery to identify degraded lands, select native tree species for restoration, and generate revenue via carbon credits and timber, scaling forest restoration in Brazil's Amazon and Atlantic forests.[16] In Kenya, AstraZeneca's AZ Forest programme is planting 6 million trees across six counties while using AI and deep learning tools to monitor progress.[17] By analysing satellite and drone imagery, these technologies track tree survival, assess health and estimate carbon capture, ensuring that restoration efforts are scientifically robust, cost-effective and climate-resilient.

AI FOR COMPLEX CLIMATE FRONTIERS

As climate change intensifies, its impacts are being felt across some of the world's most fragile systems including coastal zones, glaciers, permafrost regions, biodiversity hotspots, and vulnerable communities. AI can now play a critical role in helping us understand and respond to these evolving risks. In coastal cities, AI can be leveraged to analyse wave patterns, sediment shifts, and tidal flows to guide decisions on where to reinforce sea defenses, restore mangroves, or reroute critical infrastructure. AI can be used to monitor glacier retreat and anticipate downstream risks such as changes in freshwater availability or the threat of sudden glacial lake outbursts. In Arctic and mountainous regions, AI can help detect permafrost melt and enable early warning systems for land instability, potential release of harmful viruses and pathogens that have laid dormant and frozen, and infrastructure vulnerability. As climate-related disasters grow more frequent, AI can assist in optimizing emergency resource allocation by identifying high-risk neighbourhoods and recommending strategic locations for relief operations.

Governments may also use AI to forecast climate-induced migration by combining environmental and socioeconomic data, allowing them to plan better cities, but also undertake relocations and rehabilitations in advance. In the realm of biodiversity, AI tools can help ecologists simulate habitat shifts and species loss, supporting the design of conservation zones and ecological corridors.

AI, therefore, can become a key enabler of comprehensive climate resilience, offering new pathways for protecting people, ecosystems and infrastructure in an increasingly volatile world.

THE VICIOUS CYCLE

As if using the term dozens of times wasn't enough, we now have an entire subsection dedicated to it.

What you have read so far in this chapter represents just a few of the potentially thousands of ways in which AI can aid climate adaptation and help build climate resilience. Besides the protection of lives, ecosystems, infrastructure and economy, there is another core reason why AI must be used as a core instrument to help with adaptation and resilience: the pressing need to break a *vicious cycle* that is at play here.

Let us understand this with one or two examples. Forests act as natural carbon sinks. Trees absorb CO_2 from the atmosphere and store it in their trunks, branches, leaves and roots. Extreme heat conditions, triggered by climate change, can lead to wildfires that can burn down acres of forests. Besides releasing tons and tons of carbon as they burn, we also end up losing a massive natural carbon reserve. Moreover, the government is likely to use the funds that were earmarked for more afforestation initiatives for immediate rescue and relief efforts, and building back hundreds or thousands of destroyed houses. The forest is now not only devoid of trees, but funds that were meant to increase tree cover are now scarce too. Because of this loss of natural tree cover, the soil on the slopes comes loose, and rainfall can cause landslides, leading to even more damage.

In terms of food security, floods and drought triggered by climate change can destroy crops. Destruction of crops makes food scarcer, potentially unleashing violence in fragile societies. This is what happened in Asha's village. As violence and conflict escalate, finance gets channelized into fighting it, and rebuilding cities and towns hit by it. It then becomes difficult to invest that money into conflict hit areas to improve agriculture. Therefore, what we have at play here—and what we have touched upon repeatedly—is the existence of a *vicious cycle*. We need to break out of it and towards that end, AI offers much promise and hope.

THE POWER OF DATA INTEGRATION

Most of us use ride-hailing apps such as Uber. It may appear effortless: just set your pick-up location, get matched with a driver and wait for the car to arrive.

But in reality, there is a lot that goes in the background. These apps work by bringing together different sets of data such as your location, traffic conditions, driver availability, mode of digital payments, routes with chargers if the cab you are opting for is an EV. When combined, this data helps the app quickly match you with a nearby driver, choose the fastest route, and give you an accurate fare and arrival time. It feels simple, but it all hinges on smart data integration. Imagine if all of these diverse data points were not combined. You would then go back to the good old days of calling up a taxi company to book your ride (some of you reading this are unlikely to be familiar with those times or experiences) or whistling and stopping one whizzing past you (which, admittedly, can be easier but the fares are exorbitant).

This is the case with AI as well. Used on separate sets of data in a siloed manner, AI will absolutely produce powerful results. Applied to diverse and integrated data points, AI will not only produce powerful results, but it would be able to see how those data points relate to each other, how one parameter affects the other, and ultimately generate more inclusive and impactful results and insights. Let us take our town of Climaville. Perhaps, it faces seasonal floods, heatwaves, and rising cases of climate-sensitive diseases. With the escalating impacts of climate change, such emergencies can become increasingly foreseeable realities. AI models could begin by forecasting an approaching cyclone using satellite imagery and climate data. At the same time, public health dashboards combine information on hospital capacity, population density, and known hotspots for disease transmission. Infrastructure data, including the current condition of roads, bridges and drainage systems, informs the system about which evacuation routes remain safe and operational. This enables AI to guide emergency teams on the fastest and safest paths for moving people and delivering essential supplies such as oxygen tanks, vaccines, clean water and food. This level of integration enables not only an effective emergency response but also long-term resilience planning. It ensures that infrastructure, healthcare and logistics systems remain adaptive and responsive under pressure. AI, when powered by interconnected data sources, becomes the operational brain of a city. It processes complex information, anticipates cascading impacts, and coordinates decisions that protect lives and livelihoods.

Just outside the town of Climaville lie massive agricultural fields where farmers grow crops. By merging seasonal climate forecasts with crop disease models and real-time market data, AI can provide farmers with specific advice on what to grow, when to plant, and how to access the most rewarding markets. This could empower farmers to adapt to changing conditions while strengthening food security and economic stability.

BEYOND ADAPTATION AND RESILIENCE

Among possibly thousands of AI use cases for climate adaptation and resilience—some of which we have touched upon in this chapter—AI can truly act as a force multiplier and bring about population-scale impact. It can help save lives and livelihoods, and it can help us better adapt to the realities of a warming world.

To unlock AI's full potential for climate resilience, we must combine it with diverse, openly shared data, foster collaboration over competition, and ensure solutions are grounded in local realities. Strong political and industry will, clear policies and sustained investment are essential to turn ambition into measurable outcomes, and we will explore these aspects in greater depth in subsequent chapters.

Yet, as climate change continues to unfold, we must ask ourselves: can AI go beyond helping us adapt and build resilience to also address the root causes? Can it help us move from simply managing the consequences to actively preventing climate change itself?

10

DECARBONIZING THE PLANET

AI for Climate Mitigation

Before the flash floods, Climaville was an unremarkable city, nestled in a valley. The hills that surrounded the city helped keep the summers warm, preventing them from getting unbearably hot. The winters were crisp, but not harsh. The river flowing down the valley made winters misty, giving the city an almost tranquil feel.

As one moved away from the compact city centre towards the hills, tall buildings and busy streets gave way to quieter neighbourhoods, which is where Ori lived. His university was located just outside the city, close enough for convenience and far enough for peace. Driving down from the city to the outskirts, one could spot tree-lined streets, small parks and corner shops with organic produce, while billboards on the sides of the highway carried promises from mayoral hopefuls. The incumbent candidate pledged to cut crime and emissions, whilst the opponent pledged to replace congestion zones with what he called freedom lanes, with the caption 'drive baby, drive', perhaps echoing a world leader's famous line 'drill, baby, drill.' Advertisements promoting rooftop solar panels dotted the skyline too, lost among the noise of political slogans.

As temperatures soared and summers got hotter, people turned on their air conditioners, putting greater strain on the electricity grid than it was designed to take on. A few years before, opening windows was enough. As temperatures rose, first came the fans, and then the ACs. One day, the electricity grid just collapsed, and the city started to resort to loadshedding.[1] The demand for portable generators soared. Residents queued up at gas stations to get diesel for their generators, driving there in their pickup trucks running on diesel and petrol, causing congestion and traffic snarls, adding even more to emissions.

A year before the floods, the situation reached its peak. The mayor called an emergency meeting and it was decided to set up a new 'Department for Frontier Technologies and Climate Change Management (DFTCCM)'. Ori, who at the time was in the third year of his university, decided to intern with this office. Within the first few months, and before the summer started to creep in, a number of forward-looking initiatives were announced. The local power agency had partnered with a climate-tech startup to pilot an AI system that forecast demand and managed supply in real time. Ori had an important role to play. He was able to convince his colleagues that startups must be given a chance, and the convention of putting out expensive tenders that only big corporations had the means to win had to be changed. The new AI system mapped energy sources including renewables, layered in weather forecasts, event calendars and past usage, and built a digital twin of the grid. This allowed it to predict when and where demand would spike, stagger charging time for the city's new electric buses, and send nudges to households to run heavy appliances at off-peak times (i.e. intervals of comparatively low load on the electrical grid, resulting in lower overall demand). Solar rooftop panels were synchronized citywide, with AI showing the storage capacity needed for nighttime demand. An app even showed residents their potential solar generation and real-time savings using satellite imagery. Many signed up immediately, and by summer, the power situation started to crawl back to normal. The demand for solar panels slowly started going up at the same time. Traffic, another headache, also began to ease by the end of the year. Instead of rigid timers, AI-powered traffic lights drew on data from smart cameras and commuting apps to adjust signals based on queue lengths, speed and weather. Within months, vehicle idling times fell by a quarter. People still drove, but congestion thinned, emissions fell and air grew cleaner. Quiet improvements made a visible difference to the city.

These interventions showed that decarbonization does not always need radical reinvention. Often, it needs smarter systems. AI not only optimized what already existed but also pointed towards transformative solutions that could reshape cities: intelligent public transport, energy-efficient districts and sustainable logistics. In this sense, it acted as a force multiplier, converting ambitious climate ideas into tangible progress.

But if Climaville was making all the right moves, why did the flash floods still devastate it? The answer is simple: climate is a *global commons*. The term refers to a shared system that transcends national boundaries and is governed by no single actor. Think of outer space. Now, extend this to climate. Everyone suffers the effect, and a single party cannot stop it. Just because one city reaches net zero does not protect it from warming caused elsewhere. Does that mean efforts are futile? Not at all. Climaville's story underscores a larger truth. If the world acted with the same determination together, the impact would be profound. No change will happen overnight. But over time, collective action would lay the foundations for a safer, more resilient future for all of humanity. Certainly, climate change does not stop at borders. But if every country decided to go climate positive, climate change can be stopped across them.

With this optimistic thought, let us now explore how AI can help cities like Climaville decarbonize.

CHARGING AHEAD: TRANSPORT AND LOGISTICS REIMAGINED

Globally, the transport sector is responsible for nearly a quarter of energy-related CO_2 emissions. This includes emissions from cars, trucks, trains, airplanes and ships. Some of these segments, such as shipping, heavy-duty freight and aviation, are especially difficult to decarbonize and are often referred to as 'hard to abate' sectors. While AI can help reduce emissions across the board within the transport sector, its impact can be especially transformative in some areas.

Take urban congestion, for example. Every extra minute a car spends idling in traffic burns more fuel and releases more emissions. A journey that should take twenty minutes but stretches to fifty makes your fuel consumption and carbon footprint more than double. This is where AI steps in. By analysing real-time traffic data, it helps city planners understand how congestion builds, not just at one junction but across an entire network. It can model how a blockage at one signal ripples through surrounding intersections and creates citywide bottlenecks. If you have ever used a navigation app and noticed roads turning red when they are congested, that is real-time traffic sensing at work. Now imagine scaling that same intelligence to power a citywide traffic system. When combined with data from weather forecasts, festival calendars, sporting events, construction alerts and rainfall predictions, AI can help planners take preemptive steps, reroute traffic ahead of time, send advisories or adjust signal timings automatically across the network. More importantly, in moments of unexpected disruption such as an accident, flash flood, or emergency, these systems can respond in real time to ease pressure on the roads. When paired with adaptive traffic lights (imagine a scenario where all traffic signals are centrally controlled, with AI powering their intervals in real time), city planners are able to not just monitor the situation, but also manage it actively, and in real time. The result is smoother traffic, shorter travel times and significantly lower fuel use. The city of Pittsburgh deployed an AI-powered traffic management system called SURTRAC, developed by Carnegie Mellon University. Using real-time data from sensors and cameras, the AI adjusts traffic signals dynamically, reducing travel times by 25 per cent, idling by 40 per cent and emissions by over 20 per cent.[2]

There are hundreds, if not thousands of use cases of AI across the transport sector, especially within the realm of urban mobility. One particularly powerful area is its role in accelerating the shift to electric vehicles or EVs. While EVs are central to reducing transport emissions, one of the biggest barriers to their widespread adoption remains the lack of accessible, reliable charging infrastructure.

AI can help city planners identify exactly where to place charging stations to maximize usage and convenience. By analysing data on traffic flows, driving patterns, land use, power availability, grid capacity, and real estate availability, AI can map areas where demand for charging is likely to grow. It can also factor in future projections, such as population density trends or the expansion of ride-sharing fleets. This allows governments and utilities to make targeted investments that actually meet the needs of drivers, rather than relying on guesswork.

AI also offers transformative potential for the logistics sector, which is a major contributor to emissions. The sector is also particularly tough to decarbonize given the size of the trucks (the battery sizes are, in a way, proportional to the size of the vehicle; it is not always feasible to manufacture such large battery packs for heavy trucks loaded with heavy goods traversing thousands of miles). However, even as electrified highways with a dedicated lane for trucks and green hydrogen are being considered, an immediate fix would revolve around ensuring that better optimization takes place. For instance, Uber Freight's AI platform dynamically matches freight capacity with demand, minimizing empty truck miles, also called deadhead miles. These are the miles trucks travel with no cargo, wasting fuel and adding unnecessary emissions. Since launching in 2023, this platform has cut such empty miles by 10–15 per cent, leading to a corresponding reduction in emissions and a positive impact on sustainability.[3]

SMART GRIDS AND FORECASTING: BALANCING POWER IN REAL TIME

The power we get from the electricity grid is contributed to it by different sources. A challenge with some renewable energy sources is the fact that they are intermittent in nature. Unlike coal, or even nuclear power plants that generate energy 24/7, this may not always be the case with wind and solar power. The sun sets at night, so solar production drops to zero. On rainy days, solar will not generate much power either. Wind generation is intermittent too. The ambitious solution to this is to have large energy storage facilities (think of them as mega battery packs) where such energy can be stored, and then supplied even when the renewables are not generating power (remember the battery storage and gravity energy storage solutions we mentioned in Chapter 7?). At the same time, when a grid is powered by a mix of sources—both clean and non-clean—we need to know when and how we should use the energy. This means we use our heavy appliances when the mix of power in our network is majorly green and clean.

AI is rapidly becoming the grid's best assistant. WattTime, which started at UC Berkeley, uses an AI-powered software called Automated Emissions Reduction (AER) to help users lower their carbon footprint. When you turn on an appliance, you usually have no idea whether the electricity powering it is coming from a polluting source like coal or a cleaner source like solar or wind energy. But if you knew that, you could choose when to use electricity more wisely. This is exactly what AI-enabled interventions such as WattTime could provide. Their software can connect with smart devices and systems to shift the use of appliances to times when the energy being supplied is cleaner. For example, your dryer could automatically start running when there is extra wind or solar power available. The same idea applies to large industrial users, who can use these tools to make their operations and purchases more sustainable. WattTime now has global AI-powered emissions models that cover nearly all of the world's electricity use. That means billions of devices can reduce emissions just by knowing the best times to use electricity. On a larger scale, this also helps people investing in clean energy decide where to build new projects for maximum impact. It makes more sense to build solar or wind farms in areas where the grid still relies heavily on coal, rather than in places that already have a lot of renewable power. That way, every dollar invested replaces more dirty energy and cuts more emissions.

This also helps companies make smarter choices about where and when to get their electricity. For example, if a company was getting its power from City X, where electricity is usually dirty (i.e., generated from higher-emission or fossil-fuel sources) between 4 p.m. to 9 p.m., it could consider switching to City Y, which gets cleaner power at the same time, or it could simply shift its operations to cleaner hours in City X. But to do this, companies need better visibility on where their power is coming from, and when it is clean. That kind of insight is enhanced when AI pulls together and analyses many different kinds of data.

AI is also helping individuals and cities unlock the potential of solar energy. One example is Google's Project Sunroof, which uses satellite imagery, 3D mapping and AI to estimate how much sunlight hits rooftops throughout the year.[4] Most people do not know whether their roof is suitable for solar panels, and this lack of clarity often delays decisions such as whether to go for installation of rooftop solar, or not. Project Sunroof provides a personalized report showing how much solar energy a rooftop can generate, how much money it can save, and how many kilograms of carbon emissions it could prevent annually. AI analyses roof shape, nearby shading, seasonal sun angles and weather patterns, then combines this with electricity costs and solar incentives. This helps homeowners make informed, confident decisions about installing solar. By making solar potential visible to individuals, utilities and city planners, such AI-based solutions can encourage greater solar adoption, helping cut down emissions.

GREEN HYDROGEN AND CARBON CAPTURE: BUILDING THE NEXT FRONTIER

Green Hydrogen (GH_2) and Carbon Capture have been emerging as promising technologies to contribute to emissions reductions at a mega scale. You may recall that green hydrogen can serve as a clean fuel and play a critical role in decarbonizing hard-to-abate sectors such as cement, fertilizers and steel. Carbon capture, in parallel, can intercept carbon emissions from these industries and either convert them into usable products or store them safely underground.

While both these technologies are extremely powerful tools in our ambition to decarbonize our world, they require a complex set up. It is not as simple as installing pre-fabricated solar panels where there is ample sunlight and connecting them all together. Technologies like carbon capture and GH_2 require precise control over a number of parameters, and they are also relatively expensive. Cutting small inefficiencies can help bring down costs and accelerate their deployment. This is where AI plays a powerful role. In the case of GH_2, AI can optimize the performance of electrolyzers (which split water into hydrogen and oxygen) by continuously adjusting operating conditions based on inputs from various sensors. It can also create a digital twin of the entire production process and allow for simulation of 'what if scenarios' while also helping undertake predictive maintenance. Moreover, AI can help in enhancing energy efficiency by aligning hydrogen production with periods when renewable electricity is cheapest and cleanest (remember, it is renewable energy that is used to split water and extract hydrogen, that is why it is 'green'). AI also helps in identifying the most efficient locations for such facilities, and forecasts demand and usage patterns, making storage and distribution of hydrogen more responsive and economical.

Siemens, for instance, is offering a number of AI-enabled tools, including genAI enabled chatbots that help with better planning of GH_2 production plants and help manage complex workflows during the production cycles. Hydrogen normally exists as a gas, but storage in the liquid form is safer and economical. In the United States, Argonne National labs were able to use AI to analyse 160 billion molecules and identify forty-one that could serve as hydrogen carriers in a liquid form in just 14 hours, something that would have taken a normal computational model up to 5 years to achieve.[5]

For carbon capture, AI helps make the process of trapping CO_2 much more efficient. It does this by constantly studying live data and tweaking chemical compositions that are used to absorb the CO_2, so that more carbon is captured while using less energy. AI is also useful in figuring out the best places to set up carbon capture units, such as which power plants or factories would lead to the biggest climate benefit. It can also predict how much CO_2 will be released at different times and can help optimize the supply chain for the captured carbon. All of this makes carbon capture easier, cheaper and more likely to be used at scale.

In the United States, Microsoft and Pacific Northwest National Laboratory (PNNL) were able to sift through vast chemical libraries and databases, and ended up discovering—in just 200 hours—a new compound that could help capture CO_2,[6] which would otherwise have taken months, or even years. In the UK, Orbital has been able to use AI to innovate a sorbent that can absorb more CO_2 at higher temperatures from data centres through Direct Air Capture (DAC), with the possibility of the cost going down significantly.[7] A number of companies are also using AI-powered simulations to finetune how CO_2 is captured by dynamically adjusting the mix of chemicals used to trap CO_2 based on real-time feedback, so they capture more gas while using less energy.

DATA CENTRES AND DIGITAL CARBON

It wouldn't be a stretch to say that the world today runs on data centres. While they do not supply electricity, they provide the technology that underpins almost everything we do. From streaming videos and OTT content, to using generative AI for answers, to running complex softwares that design factories or manage global supply chains, to our email inboxes and every video call we make, data centres are the giant brains powering and processing it all. But this digital backbone comes at a steep cost. Data centres consume enormous amounts of electricity and water, leaving a significant carbon footprint. Today, they use as much cumulative power as the entire country of France, and by 2030 their consumption could rival that of India. This reveals a paradox at the heart of AI: while it holds immense potential to fight climate change, it also contributes to it. We discuss this tension in the chapters ahead, exploring this challenge and its solutions. For now, it is enough to note that AI, powered by data centres, can also be deployed to help decarbonize those very data centres.

AI can help optimize the water flow to data centres. It can optimize the cooling systems to run in the most efficient manner possible, while drawing a minimum amount of power. It can also be used to better synchronize power consumption with the availability of renewable energy. Using GIS and advanced analytics, it can also suggest where co-locating data centres with solar, wind and battery storage infrastructure would be most efficient. It can help balance workloads across servers, ensuring efficient resource use and reducing unnecessary energy draw. By predicting equipment failures early, AI enables preventive maintenance, minimizing downtime and energy waste. It can also create digital twins or virtual replicas of data centre systems to simulate and optimize performance without physical trials. AI can also assist in aligning computing tasks with times when renewable energy is available, a method known as carbon-aware computing. Additionally, it can shift workloads geographically to tap into cleaner power sources in different regions, further reducing emissions.

Meta has partnered with Amrize and the University of Illinois to build data centres using AI-optimized concrete mixes. Using AI, they refine concrete formulations to meet data centre requirements, balancing strength, durability, thermal regulation and energy efficiency. This approach is estimated to cut the carbon footprint of the concrete by about 35 per cent, which in turn helps reduce the total emissions of the data centre itself.[8] It is a strong example of how AI can lower both direct and indirect carbon emissions, the latter coming from the choice of concrete and the way it was being produced. The indirect and direct emissions is a point worth remembering. We will return to it towards the end of this chapter. Google has used DeepMind's AI to improve the energy efficiency of its data centres by managing the cooling systems. The AI dynamically adjusts fans and chillers in real time, leading to a 40 per cent reduction in cooling energy use and a 15 per cent decrease in overall energy consumption across its data centre operations.[9]

MATERIAL INTELLIGENCE: CEMENT, STEEL AND THE HARD-TO-ABATE

As mentioned earlier, it is often extremely tough to completely eliminate emissions in some sectors, at least in the immediate term. These hard-to-abate sectors include industries such as cement and steel and transport modes such as shipping, aviation and heavy-duty trucks. Combined, these sectors account for over 25 per cent carbon emissions, globally.[10] Moreover, these emissions will only increase as the world population increases and cities rapidly urbanize. Buildings, roads, flyovers, railways and metro rails all require cement and steel. In cement, most emissions come from the chemical reaction that turns limestone into cement, a process that naturally releases CO_2. In steel, the traditional method of using coal to extract iron from ore also produces large amounts of emissions. Further, the melting of scrap steel is also carbon intensive if the electric arc furnaces used to melt it draw power from a dirty grid (a grid which is predominantly supplied by fossil fuels).

ArcelorMittal, one of the world's largest steel manufacturers, worked with Smart Steel Technologies to deploy an AI-powered solution to improve wire rod production.[11] Traditionally, the beginning and end portions of wire rods are trimmed off because they have lower quality, resulting in material loss and higher emissions. By analysing real-time process data alongside past production records, the AI system now pinpoints the precise moment to make the cut, reducing this 'trim scrap' by 20 per cent. This not only improves quality but also lowers electricity and gas use, leading to a measurable drop in CO_2 emissions. Carbon Re, a UK-based AI startup spun out of the University of Cambridge, is developing advanced AI tools to help decarbonize foundational materials like cement. When its software was integrated into the ABB optimization system at Heidelberg Materials' cement plant in Mokra, Czechia, the factory achieved a 2 per cent reduction in fuel-related emissions within just the first month of deployment.[12]

Even material design is being transformed. Earlier, scientists would manually test different combinations of materials, mixing various metals or cement ingredients, and adopt only a few based on years of trial and error. Today, generative AI acts like a co-designer, simulating millions of combinations to find the best ones quickly. For example, AI can help develop new types of steel that are just as strong but produce fewer emissions, or cement mixes that hold buildings together with less environmental impact.

It is similar to how AI recently identified the most promising compounds to carry green hydrogen in liquid form, not in years, but in just a few hours.

TRUST, TRACEABILITY AND TRANSPARENCY: CARBON MARKETS AND MEASUREMENT

Offsetting emissions means acknowledging your own carbon output while investing in projects elsewhere such as reforestation or renewable energy to counterbalance it. This is, broadly, the core idea behind carbon markets. If you emit 1 tonne of CO_2, you pay someone else to reduce or remove 1 tonne elsewhere. However, a major challenge arises. How can a steel plant owner in Austria be sure that the reforestation project they are funding in Colombia is actually being implemented as claimed? How do they avoid paying for reductions that are either exaggerated, duplicated or entirely fictional?

This is where AI is becoming critical. Startups like Pachama are using AI to evaluate forest carbon projects with tools such as satellite imagery and lidar.[13] Their systems independently assess the quality and durability of carbon credits, offering buyers, investors and governments a clearer picture of whether their offsets are achieving the intended climate benefits. These AI-driven tools not only help verify that carbon reductions are real and occurring in the right places, they also build trust in the process. By making carbon reduction efforts more transparent and traceable, such technologies give current and potential investors confidence that their contributions are both meaningful and measurable. Another powerful example is Climate TRACE, which uses satellite imagery, remote sensing and artificial intelligence to monitor and map greenhouse gas emissions from sources around the world, including power plants, steel mills, oil refineries and ships.[14]

This AI-powered system makes emissions not just traceable but also predictable. It allows for a far more detailed and near real-time view of where emissions are coming from and how they are changing. Such intelligence enables governments, companies and investors to deploy mitigation and offsetting strategies more effectively, focusing on locations where interventions will have the greatest climate impact, regardless of national borders. It can also inform corporate procurement decisions by revealing the carbon footprint of materials, supply chains and the geographies involved. Crucially, by making this data publicly accessible, Climate TRACE promotes transparency and accountability. Such solutions offer a strong alternative to outdated paper-based inventories and provide the world with a more honest view of what is happening on the ground. Such interventions can be thought of as foundational steps for achieving a credible, effective and equitable global transition to net zero.

AGRICULTURE, AND MORE WAYS AI CUTS EMISSIONS

While earlier sections explored detailed AI applications in transport, cement, steel, carbon traceability and energy, these represent only a small slice of the full picture. The number of AI use cases that directly or indirectly reduce emissions runs into the thousands, spanning every major sector. To document them all would require an entire volume of books. This section offers a glimpse into the many other areas where AI can help in cutting emissions, but again, this is not exhaustive.

Agriculture is a critical frontier. It contributes nearly a quarter of global greenhouse gas emissions, driven largely by methane from livestock and nitrous oxide from fertilizers. In the dairy sector, AI tools are being used to track the health, diet and digestion of individual animals, and identifying which cows emit more methane and why. Based on this data, farmers can adjust feed types, improve breeding programmes and reduce methane output. AI can also help monitor and contain methane emissions from manure pits and biogas systems. These tools can also alert operators in real time if there are leaks, helping prevent atmospheric release of methane, which is far more potent than CO_2. AI is also helping reduce fertilizer-related emissions. These primarily come from the use of nitrous oxide, one of the most powerful greenhouse gases. Instead of applying nutrients uniformly across large fields, AI can recommend targeted application based on soil condition, crop growth stage and weather forecasts. These precise insights can help reduce the corresponding emissions. These insights are not only useful for managing fertilizers but also help save water used for farming. They do this by suggesting better ways to level the ground and apply water more efficiently, so less water is wasted while keeping crops healthy. This reduces groundwater overuse and the emissions that are generated through agricultural pumps that run on electricity.

Food waste is another hidden emissions driver. When food is discarded and decomposes in landfills, it releases methane. AI helps predict demand more accurately across supply chains, allowing grocers, suppliers and consumers to minimize overproduction. Smart inventory and pricing tools powered by AI also help move food before it expires, ensuring that less is wasted and less ends up rotting in dumps. Again, the point to remember here is how AI can cut direct and indirect emissions across the entire supply chain. In cities, AI helps with management of waste collection, segregation and recycling as well as managing the water supply efficiently. AI can be used to monitor urban water systems, detect leaks, improve pumping efficiency and guide treatment systems. AI can also help in managing air quality, which you read about in the previous chapter. In the process of improving air quality, the corrective measures undertaken would also involve those that cut down emissions. These measures could be congestion pricing, transition to EVs, building metro systems that are powered by renewable energy and so on. Even in how we think about materials and waste, AI plays a role. By predicting when items like electronics or machinery will reach end of their life cycle and identifying where these materials are available, AI can help match them to new uses or recycling pathways. This reduces the need to produce new goods from raw materials, which often involves high emissions and resource use.

These interventions might seem small in isolation but together they create a system that is more efficient, responsive and low-carbon. They also highlight that the power of AI lies not only in big industrial revolutions but in solving thousands of small, scattered problems that were too complex to handle before. AI connects the dots in ways that lets us reduce emissions before they accumulate, giving us a fighting chance at climate action where it matters most.

SCOPE 1, 2, 3, DATA INTEGRATION AND DIGITAL TWINS

What we have seen across these examples is the number of interdependencies between various sectors and how emissions can be cut directly, but also indirectly. For instance, cement production is a tough sector to decarbonize, but the same cement is also used to build data centres, whose increasing energy consumption has people worried. AI can help cut emissions in data centres, in the production of cement, and in identifying mixes of cement that, if used to build data centres, would help keep them inherently cooler, further reducing energy demand. AI can also predict where large-scale solar farms should be built, and can throw out insights that could lead to data centres and solar farms being co-located.

AI can also suggest how compute loads should be divided across data centres based on which centre is running on the cleanest power at a particular time of day.

So, for someone who is building a data centre, their contribution to emissions reduction is not just how much of their own energy is green, but also whether the materials they used to build the data centres are green (i.e. how much was the carbon that was emitted when the steel and cement that was used to build this data centre was produced). Moreover, when this material was transported by a logistics company to the data centre, what were the emissions of the huge trucks that brought them there? This is perhaps the time to introduce Scope 1, 2 and 3 emissions.

Scope 1 covers direct emissions. If a data centre uses diesel generators, the emissions from burning diesel are counted as Scope 1.

Scope 2 involves indirect emissions from the energy the data centre consumes. For example, if the grid supplies coal-based power, the carbon footprint associated with generating and supplying that electricity counts as Scope 2.

Scope 3 includes all other indirect emissions along the supply chain, such as the manufacturing of equipment, the use of materials like cement, and transportation.

For instance, emissions from the production of cement used in building the facility would be Scope 1 for the cement manufacturer, while the emissions from transporting materials to the site or the energy used to produce the cement would be Scope 3 for the company that owns the data centre.

Even from the limited set of AI applications, it's clear that technology has the potential to reduce emissions across three scopes, cutting both direct and indirect emissions and helping organizations become more sustainable.

The other crucial point, as emphasized in the previous chapter, is the extraordinary power of integrating data across all these use cases. Think about it. Data from data centres, the cement used in constructing them, the grids that power them, and the timing of when that power is clean or carbon-heavy. Add to that the energy consumption patterns of billions of people, their transport and food choices, the resulting emissions, insights from satellite imagery, and the full picture of where emissions come from. Consider also the products we buy, where they are made, what they emit, the flow of traffic on our roads, the heat profiles of our buildings, and the material options available to cool them more efficiently. Now multiply that by billions of data points. When these fragments of information are integrated and analysed through AI, the possibilities become transformative, delivering insights and solutions that would have been unimaginable just a few years ago.

OUT OF THE (VICIOUS) LOOP, INTO THE FUTURE

A prominent and possibly irritating theme, given how often we return to it, is the *vicious cycle* we have described multiple times by now. Much like in the previous chapter, a dangerous loop exists in climate mitigation as well. For instance, rising emissions cause climate change, which then leads to flooding events in cities. These flooding events can damage infrastructure across the city, including infrastructure that was aiding mitigation such as smart traffic sensors or energy storage sites. This is what happened in Climaville. Another example is the growing risk of wildfires and floods, worsened by climate change. These disasters can damage renewable energy infrastructure such as solar farms or power lines, or destroy forests that were planted to offset carbon. Immediate recovery and rebuilding will assume priority, and investments which could have been made to deepen mitigation and adaptation, now get pumped into those efforts instead. Not only is progress undone, but the path forward becomes harder, costlier and carbon-intensive yet again. Thus, a reinforcing loop keeps setting us back just when we need to move forward.

Can AI break this loop? It potentially can, but it needs policy, finance and a ton of other levers to back it up.

With the right levers, AI can do much more than just aid mitigation and adaptation. It has the potential to transform how we manage climate change on a systemic and institutional level, ensuring solutions are sustainable and deeply integrated across society.

Despite bold strides in emission reductions and resilience, many cities face hurdles in scaling their impact without better data, transparency and coordination. By harnessing AI not only to manage resources but also to guide choices, shape policies and connect sectors, cities can move beyond being climate-smart towards becoming truly climate-intelligent.

The next chapter examines how AI's power can be unlocked to transform climate governance and drive system-wide climate action.

11

INTELLIGENCE AT SCALE
AI for Systems Transformation and Climate Governance

They say the world is small, and that is certainly true to some extent. Do you remember Asha's uncle from Chapter 3? Her uncle had moved abroad and would often video call Asha and show her his city. He was poised to run for political office. He did, and is now the current mayor of Climaville, Mayor Ashman.

It was Mayor Ashman who set up the Department for Frontier Technologies and Climate Change Management (DFTCCM) which had hired Ori as one of their first interns. It was under him that the city had started mainstreaming AI for climate mitigation, and it was under his watch that the city had witnessed the terrible flash floods. He has been working in mission mode, coordinating relief and rescue efforts across the city. The flooding has provided him with proof that AI must be mainstreamed for adaptation, for building resilient infrastructure, for setting up early warning systems, and for ensuring that even in case of extreme weather events, people and infrastructure, to the extent possible, stayed safe.

His DFTCCM had already contacted experts, and a plan of action was being prepared to integrate AI-enabled solutions that could predict the unpredictable.

And so, in Climaville, action is already underway. New data centres are being built on higher ground, safely out of reach from rising waters. City officials have opened up key city-related data points to the public, and a digital twin of the entire city is now in play, helping planners spot weak points and act before disaster strikes. They have already run simulations showing that the next big flood could damage two major bridges, so reinforcement work has begun. Meanwhile, an early warning system powered by AI is watching everything in real time. It is plugged into thousands of sensors, drawing data from electricity grids, traffic signals, air quality monitors and water level gauges. It learns from current signals and real-time inputs while digging through historical records of hundreds of past events, giving the city a real chance to stay one step ahead.

Climaville had already been working hard on mitigation for a while, and we have seen in the previous chapter how AI and GIS-guided solar panels have been installed strategically, energy storage capacity has been brought in, and heavy appliances in houses and high power-consuming machinery in factories run when the power in the grid is the cleanest. AI has also helped better manage traffic, and congestion at traffic signals across the city has reduced, bringing emissions down significantly. Some of their mitigation infrastructure was destroyed by the floods, and Mayor Ashman faces a tough choice in allocating funds; he must prioritize rehabilitation first. It will take time, but they will be able to build back. (Now imagine the difficult decisions policymakers in developing countries must face, often with far lesser resources and support). The city is much better off than many neighbouring ones, especially Carbonville and Oilville, and Climaville takes immense pride in being at the forefront of climate innovation.

Yet, despite the many AI-powered initiatives that they have taken on the mitigation and adaptation front, and their strong ambition, they have still not been able to keep pace with the scale of the climate challenge. Mayor Ashman had pledged to make Climaville net zero by 2040, but progress was faltering. Emissions data arrived months too late. Investors were hesitant to fund future projects without better visibility into risks and returns. Citizens were extremely eager to play their part, but there was either too little information that could help them make a sustainable choice, or there was too much of it, which made them confused. Turning to his latest hire, Ori, Mayor Ashman decided that he would turn to AI for a solution. Through a pilot project, Climaville used AI to assess the impact of its climate spending across departments.

The system flagged gaps where funding had not translated into emissions reduction. They also realized that the intelligence generated by AI, grounded in real-world data, told a different story than what was mentioned in various government reports, which often portrayed a more optimistic picture than the actual situation on the ground. Mayor Ashman also established a Behavioral Insight Unit that combined AI with data on spending and choice patterns to design effective nudging strategies for citizens. The unit helped suggest simple daily actions such as switching to sustainable packaging or changing commuting habits, and showed users the emissions they avoided through a dedicated app. In partnership with a popular search engine, anyone searching online for petrol vehicles was also shown electric alternatives, complete with personalized comparisons of cost and emissions based on their usage patterns.

Noticing the promising potential of AI in transforming systems, bringing transparency and improving climate governance,

Mayor Ashman also decided to introduce a policy that would encourage all government departments to leverage AI for optimizing, tracking and monitoring investments made in climate projects and advancing market interventions and instruments such as carbon credits and green bonds. What Mayor Ashman was trying to do was interesting: besides leveraging AI for mitigation and adaptation, he was now interested in using it to guide everyday choices, craft better climate policies, pull in more funding, and connect the dots across sectors to drive meaningful, large-scale progress on climate action.

His vision, as ours should be, was to build not just a climate-smart city but a climate-intelligent one, driven by coordinated, AI-supported, system-wide change.

Let us now look at some of the ways in which policymakers like Mayor Ashman could bring about this change.

CLIMATE FINANCE AND INVESTMENT INTELLIGENCE

Mobilizing finance remains one of the biggest challenges in scaling climate action. Despite the billions pledged, too little flows to high-impact projects, particularly in emerging economies. Risk perception, lack of transparency, incomplete or skewed information and fragmented data are key barriers. Another challenge, or rather a market failure, is that climate financing is not always optimal. In simple words, as we have seen earlier, it does not make sense to add yet another clean energy project to a relatively clean electricity grid, whereas the same clean energy project on a dirty grid can have a much bigger impact by replacing fossil fuels. A project cannot be undertaken without finance, and therefore, all else equal, climate finance should flow into geographies or sectors where the impact of those investments will bring about the maximum change on the climate front.

Take a simple example. A firm with a lot of money is interested in supporting rooftop solar systems. They narrow down on Climaville, which now draws 80 per cent of its power from clean sources and about 20 per cent from coal. However, its neighbouring town, Carbonville, draws only 20 per cent from renewables and about 80 per cent from coal. Surely, the solar panels will have a greater impact on carbon emissions in Carbonville as they will reduce some of the demand for dirty energy supplied by coal. Therefore, it might make more sense to do this project, much to Mayor Ashman's dismay, in Carbonville. But this is only possible when we have intelligence, rich intelligence and thousands of data points brought together to show simple, clear insights to someone sitting across the table, ready to sign a cheque.

AI is helping address such gaps. By analysing diverse datasets, everything from satellite imagery and policy trends to financial flows and climate risk indices and even historical records and patterns, AI can pinpoint where finance is most needed and where it can deliver the greatest emissions impact. It also helps structure innovative instruments like climate bonds by simulating return-risk ratios under different climate scenarios.

An example of this would be Cervest's flagship platform, EarthScan, which allows users to understand how climate risks such as floods, droughts and extreme heat could affect the assets they own, manage or depend on, both now and in the future. By delivering clear and science-backed insights, solutions such as EarthScan help businesses and institutions make smarter and lower-risk decisions, meet regulatory disclosure requirements and strengthen the climate resilience of their operations and investments.[1] A lot of global action is now centred on carbon credits as governments and companies look for credible ways to meet their climate commitments. At their core, carbon credits represent the reduction or removal of 1 tonne of CO_2 or its equivalent, often earned through projects like reforestation, renewable energy, or methane capture. Companies buy these credits to compensate for emissions they cannot yet eliminate. For this system to work, each credit must be real, measurable and counted only once, making rigorous verification and trustworthy registries essential. Today, AI can help connect different parts of the climate finance system, such as carbon trading markets, registries (official systems that track and manage carbon credits to ensure they are real, unique and not counted more than once), and projects focused on environmental goals. AI is increasingly used to strengthen monitoring, transparency and confidence in carbon markets. For example, ZERO13 uses AI to manage the whole process of issuing, trading, verifying and settling carbon credits and sustainable investments digitally. It uses real-time data, smart analysis and blockchain technology to make sure everything is trustworthy and linked appropriately and seamlessly, solving common problems like lack of trust and disconnected systems in climate financing.[2]

AI FOR BETTER CLIMATE POLICY PLANNING AND GOVERNANCE

Designing climate policies is one of the most difficult challenges in governance today. It demands that governments strike a delicate balance between environmental ambition, development, economic growth, political realities and social equity. But even when the right policies exist on paper, their true impact depends on how and where they are implemented. The challenge lies in the fact that climate change touches every sector, and every sector is deeply interconnected. A policy targeting one issue in isolation, say clean energy, can fall short if it does not account for dependencies across transport, housing, industry or land use. To be truly effective, climate policies must take an ecosystem-wide view. And yet, this is far easier said than done.

Anyone who has worked in government will recognize how difficult such coordination is in practice. Sectoral silos, fragmented data, legacy systems, competing mandates and sometimes even political turf battles can stand in the way of alignment. Budget allocations may differ sharply between departments, and timelines often conflict. In the developing world, this complexity is further amplified by the weight of competing development priorities. Policymakers are not just working on climate. They are managing flood recovery, fixing roads, building water systems, responding to public grievances and delivering on promises of economic growth, all at the same time. It is easy to say that climate action should be integrated into every department's work. But that overlooks just how stretched many public officials already are. If an official in a climate-vulnerable district is spending most of their energy fighting floods or wildfires, when will they have the time to plan how to prevent the next one? Where will they find the funds? And if every season brings more disasters, when do they break out of this *vicious cycle*?

AI has the potential to help here, not by replacing human decision-making, but by clarifying and complementing it. By analysing vast amounts of cross-sectoral data, forecasting risks, identifying overlaps and mapping co-benefits, AI can support governments in designing policies that are holistic, better targeted and ultimately more impactful. But for that to happen, governments need the capacity, tools, political will and financing to use such intelligence well. In an ideal world, every city would have the foresight of Climaville, with governments led by visionary leaders like Mayor Ashman. But that is not always the case. Most governments are navigating difficult trade-offs, with limited time, data and resources. If we want climate policy to succeed, then we must support governments in making smarter choices faster, and that includes putting the best tools of intelligence in their hands.

We would suggest that AI for better climate policies and governance should serve to address three broad ambitions:

1. **Freeing Up Policymakers' Time for What Really Matters**

    Many policymakers, especially in vulnerable regions, spend a significant portion of their time firefighting and responding to citizen grievances, chasing compliance paperwork, or manually appraising and reviewing project proposals, including those that have climate implications. These are essential tasks, but they are also time-consuming and often repetitive. AI can take over many of these routine functions. For instance, AI-powered chatbots can handle a large share of citizen service requests, freeing up officials to focus on core policy work. Similarly, project appraisal, which usually means reading through long reports and risk assessments, can be made much faster and sharper using AI. These models can not only evaluate the proposal itself but also scan thousands of other data points that may not be mentioned in the report. They can draw connections with climate predictions, regional hazard models, infrastructure plans, previous project outcomes, and even the latest scientific and financial analyses. This allows for a more complete picture of whether a project is viable, what risks it might face in the future, and how it compares to other similar interventions across geographies.

    Imagine a state planning a 2 gigawatt (GW) solar park in a drought-prone district. On paper, the site looks ideal with abundant sunlight, available land and strong investor interest. But when an AI system evaluates the plan, it uncovers deeper risks that are invisible to the human eye. It finds that the district's water table has been declining for 12 consecutive years and solar panel cleaning alone could require 6 to 8 million litres of scarce water every month. Climate models indicate that dust storms will become longer and more frequent, reducing panel efficiency and doubling cleaning cycles. At the same time, grid simulations reveal that the nearest substation will face instability by 2031 unless at least 200 megawatt (MW) of storage is added. Based on these insights, the AI recommends shifting part of the project to a neighbouring block, adopting a hybrid battery and pumped hydro storage strategy, and using precision cleaning systems to save water. In seconds, AI turns a seemingly straightforward solar plan into a resilient and climate-proof energy asset. These are not things that a single official can check in a day, but AI can. In doing so, it allows limited human resources to be spent on strategic decision-making rather than administrative heavy-lifting.

2. **Sharpening Decision-Making through Smarter Insights**

AI can give policymakers the power to simulate outcomes before they act. With hybrid models that combine climate, economic and machine-learning data, planners can test how emissions, public health or jobs might respond to a certain policy, whether it is a fuel tax, a subsidy for electric vehicles or a change in land use. This makes policy not just reactive but predictive. It also helps make trade-offs transparent. It could also show that for a policy to succeed, support may be needed from other stakeholders, including different ministries or departments within the government. It can highlight how one policy might affect others or ripple across sectors, and how those stakeholders are likely to react. This allows policymakers to anticipate challenges early and design interventions that are more holistic, coordinated and inclusive.

For example, a city considering a congestion charge might use AI to forecast its impact not just on traffic, but on pollution hotspots, local business activity and public transport usage. Tools that use Natural Language Processing (NLP) to scan thousands of policy documents globally, can support this effort by letting policymakers compare what has worked in other countries. AI can also be used to simulate financing scenarios, such as testing whether a policy could be sustained under different climate stress conditions or commodity price changes. This improves both the design and durability of climate action.

3. **Making Climate Action Trackable and Responsive**

    Passing a climate law or a policy is only the beginning. Real progress depends on whether the policy is implemented as intended, and whether it continues to deliver benefits over time. AI enables real-time monitoring of progress and offers alerts when things veer off track. For example, AI systems can track energy usage patterns, air quality data and construction timelines to verify if an energy efficiency regulation is being followed. They can also help identify blind spots, such as where a policy is leading to unintended emissions shifts elsewhere, or where compliance costs are disproportionately hurting smaller businesses. More importantly, AI can guide proactive interventions. If a region is showing early signs of crop failure or power outages due to climate impacts, AI can help trigger early support or funding reallocations before the problem escalates. This is particularly important for long-term sustainability. A solution that looks great today may become irrelevant or even counterproductive in five years unless there is a system to track and adapt it. AI offers that system, not as a replacement for governance but as its support structure.

In summary, *the role of AI in policymaking is to automate and to augment*. It enables better use of limited resources, sharper insight into trade-offs, and quicker responses to change. It helps ensure that finance and governance systems are designed not just to launch climate solutions but to sustain and evolve them over time. If our cities and nations are to scale climate action at the pace and depth required, such intelligence is not optional but essential.

ACCOUNTABILITY, POLICY AND CLIMATE DIPLOMACY

At the global level, AI can be used to anticipate potential geopolitical tensions linked to climate impacts such as water scarcity, food shortages or migration patterns. By modelling climate and social data together, governments and international organizations can forecast risks and act early to prevent crises, laying the groundwork for collaborative and coordinated climate diplomacy. In an era shaped by complex and often uncertain international relations, platforms powered by AI can analyse thousands of policy documents, treaties, and climate laws. This makes it easier to compare instruments across countries, identify gaps, spot overlaps and highlight opportunities for improvement. Such tools also support better collaboration between nations by revealing where alignment is possible, while flagging areas where expanding mandates might create friction or geopolitical tension.

As companies and countries scramble to make claims about net zero, sustainability and offsets, ensuring transparency and trust is critical. AI helps verify these claims by comparing what is reported with independent data sources such as satellite imagery or supply-chain information. Algorithms can spot mismatches, flag inconsistencies in Environmental, Social and Governance (ESG) reporting, and detect potential greenwashing, helping regulators and stakeholders hold organizations accountable.

AI is beginning to support climate litigation as well. As lawsuits linked to climate harm increase, AI can assist legal teams by scanning through past rulings, environmental datasets and regulatory filings to identify patterns, locate relevant precedents and predict likely outcomes. This enables quicker and more informed legal responses to environmental violations. In the education space, AI-powered platforms are transforming how people learn about climate change. By adapting content to individual learning styles and providing interactive simulations, they help students understand complex topics like emissions, energy systems and the carbon cycle. This personalized and immersive learning experience makes climate concepts more tangible and prepares future generations to act with both knowledge and urgency.

WRAPPING UP ... OR UNWRAPPING YET ANOTHER COMPLEXITY?

This chapter explored how AI can support climate action through smarter systems, sharper governance, deeper behavioural insight, and more effective financial decision-making. We looked at how AI can simulate policy outcomes, increase transparency, direct investments to high-impact areas and support individuals in making lower-emission choices. From monitoring to long-term planning, AI's ability to integrate information across domains makes it a powerful tool for climate mitigation, adaptation and protection.

At the heart of many of these use cases is the importance of connecting data across sectors, regions and systems. The power of AI lies in drawing insights from vast and varied datasets to guide decisions that are timely, tailored and impactful. In doing so, it can help break the *vicious cycle* that climate change often creates whereby mounting damage brought about by climate change reduces our ability to prevent even greater harm.

But unlocking this potential will require a level of collaboration we have rarely seen before. Governments must recognize the role AI can play in advancing climate goals and take steps to mainstream its use across sectors. Corporates, startups, researchers and civil society must focus on building solutions that complement one another rather than compete. None of this is possible without enabling policy, smart regulation, access to climate finance and meaningful partnerships. Those advising the public sector must also focus on delivering insights that are actionable and context-aware beyond being a technically sound, yet obvious solutions.

We now understand how AI can augment climate action. The next step is to explore how we can strengthen AI, advance climate efforts and turbocharge the way they work together, i.e., their nexus. But before we turn to that, we need to acknowledge a critical tension: AI is a solution that also contributes to the problem. While it helps decarbonize the world, it also carries its own carbon footprint.

So, before we dive into the final section of the book which focuses on solutions, strategies and transformation, let us first explore some of *AI's algorithmic dilemmas*.

12

FROM DILEMMA TO DESIGN

Building Scalable, Sovereign and Sustainable AI
for Climate

Mayor Ashman was proud of himself, his government and the people of his town. The city of Climaville had endured devastating extreme weather and suffered heavy losses. But today, things were different. With the help of AI-powered systems and insights, they could now predict when extreme weather events were likely to strike. They had relocated critical infrastructure away from high-risk zones, put in place an early warning system to move people to safety in time, and set up joint government and private research centres. These centres were using advanced AI models to develop new types of cement that could resist flooding while also keeping homes cooler during heatwaves. Construction companies had already begun using these innovative materials. Across the city, essential infrastructure was being monitored around the clock, helping Climaville stay a step ahead.

He was equally proud of the progress made on cutting emissions. Some industries across Climaville had started using AI to optimize energy use and reduce waste, while a few factories had begun shifting to cleaner fuels and smarter production lines. Public transport had been upgraded with route optimization and predictive maintenance powered by AI, reducing fuel use and traffic congestion. Even food waste was being tackled through AI systems that helped vendors forecast demand more accurately and keep excess from ending up in landfills. Construction and energy use across the city was being guided by AI-informed insights and digital twins, helping reduce the city's overall carbon footprint.

On the governance front, the city had embraced AI not just to run operations better but to design better policies. Some officials were already using AI tools to simulate the outcomes of major decisions, helping them prioritize investments that would bring the biggest climate benefits. Digital platforms guided funding to projects with the highest emissions reduction potential, especially in underserved areas. AI also helped communicate with citizens, offering real-time updates and tailored nudges to shift them towards climate-friendly choices.

Ori, having graduated from university, was now a full-time employee of the Department for Frontier Technologies and Climate Change Management, or DFTCCM. He was playing a key role in Climaville's AI-based systems.

One fine morning, Mayor Ashman was particularly excited. He had just met the CEO of a major tech company that wanted to build new data centres in Climaville. Climaville did have smaller data centres, which had also been damaged in the flooding, but nothing could match the scale of what the CEO was promising him. He was thrilled at the prospect. 'We should announce this at the press briefing today,' he said, waltzing into the cafeteria with his mug of coffee. When he spotted Ori, he shared his big news. Ori smiled and nodded. It was indeed promising. After all, data centres are the backbone of AI and digital infrastructure, and this could open up new opportunities for Climaville. 'But,' he added gently, 'we should do it with a plan, and we, at the DFTCCM, have been working on it. In fact, we wanted to get time with you to discuss what we have been thinking.' Ori and his colleagues explained how not all data centres are powered by clean energy. If they run on dirty power or guzzle water in a water-stressed area, they could undo some of the city's hard-earned climate progress.

The mayor raised an eyebrow. 'But how much power do these things really use? Surely, they can't be worse than a bunch of laptops?'

Ori paused. 'Well … today, global data centres consume almost as much electricity as the entire country of France.'[1]

His eyes widened. 'Wait, what?'

'And by 2030,' Ori continued, 'if the IMF were to be believed, they could use as much power as all of India … that would be over 1500 terawatt-hours.'[2]

The mayor spilled a bit of coffee. 'How much is 1500 terra … whatever?'

'1500,000,000,000,000 …' Ori said, helpfully. 'That is fourteen zeroes, in case you're counting.'

He was.

Welcome to the algorithmic dilemma. AI's powerful algorithms can solve complex climate problems, but these algorithms are brutal when it comes to their power and water consumption, and that is not exactly ideal, is it? Especially if this power is dirty power.

As Ori walked back into his office, Mayor Ashman seemed unhappy that his press moment would no longer materialize today. However, Ori reassured him that while this dilemma was definitely a challenge, they could convert it into an opportunity.

Climaville could build one of the world's best data centre policies and attract the world's most innovative data centre designs.

Ori and DFTCCM also took this opportunity to brief him on two other things that were on the back of their minds. Firstly, there was the need to make climate data even more accessible so that solutions designed could be far more integrated, inclusive, innovative and impactful. Some developers had complained to them that a lot of extremely valuable data points were missing, or were not being collected, or were all over the place. Worse still, some of this data was put behind paywalls and private companies were charging for it. Secondly, a lot of companies, enthusiastic about jumping into the bandwagon and announcing how their AI was helping Climaville, had solutions trained on data from another part of the world. It was giving insights that were contextually irrelevant and ridiculous.

By the end of the meeting, Ori and DFTCCM colleagues were able to convince Mayor Ashman that all of these three issues—AI's power consumption, the problems associated with accessing data freely, and AI's sovereignty—were actually some of the biggest opportunities for Climaville to create a blueprint for the future of AI-based climate solutions. And this blueprint would make for a bigger press moment for him, one which would likely grab international headlines.

Just like Ori and his DFTCCM colleagues, we strongly believe that by solving the following three key and foundational challenges—

1. **AI's power consumption and carbon footprint;**
2. **data availability for building climate solutions; and**
3. **sovereignty of AI models and solutions**

—we could convert a dilemma into design, and challenges into powerful opportunities. Backed by the right policy, finance and cooperation levers, we could create a powerful blueprint for building scalable, sovereign and sustainable AI for combating climate change. Moreover, this blueprint will have a positive spillover effect across multiple sectors.

FOUNDATIONAL PROBLEM (READ: OPPORTUNITY) 1: POWER-HUNGRY AI ALGORITHMS

Imagine installing a smart thermostat in your home that uses AI to reduce your heating bill and save energy, throwing constant genAI-based insights and suggestions at you. But behind the scenes, it relies on a massive algorithm running in a distant data centre that consumes more electricity in a day than your entire apartment does in a month. In the context of climate, *AI presents an algorithmic dilemma*: while it holds immense promise for advancing green development and climate resilience, it comes with a deep carbon footprint that can no longer be ignored, especially as demand for AI shoots up, increasing the energy consumption of data centres that power it. Training large AI models requires vast computational power which consumes significant amounts of energy. If most of this energy comes from fossil fuels, then our AI is not really green or sustainable—it is generating substantial carbon emissions. The operation of data centres, which are the backbone of AI, demands enormous water resources for cooling and, in some cases, involves clearing land or forests to accommodate new infrastructure. Moreover, these centres are increasingly vulnerable to climate-related extreme weather events such as heatwaves, floods and storms.

As AI scales, so too must efforts to green it, ensuring that tools meant to fight climate change do not end up quietly fuelling it.

FOUNDATIONAL PROBLEM (READ: OPPORTUNITY) 2: DATA AVAILABILITY

One of the things we have stressed across the last few chapters is the centrality of data in developing AI-enabled climate solutions. We have seen that all sectors are deeply interlinked with each other, and so are their emissions. The way these sectors interact also determines the nature of emissions and how their clean transitions will unfold. This includes everything from direct emissions to those embedded in supply chains and energy systems, often referred to as Scope 1, 2 and 3. Moreover, the impacts of climate change do not stop at national borders. This makes cross-border data equally important for enabling tech solutions—including AI—to operate effectively at scale.

The three previous chapters have all emphasized the power of data integration. When brought together, diverse datasets can generate insights that are system-wide, inclusive, holistic and innovative. Because these insights are built from data across the entire ecosystem, they are more likely to be impactful and relevant. Imagine a city planner building a digital twin of a city that draws on real-time information from thousands of sources, as well as historical trends, and then uses that information to simulate various future scenarios and interventions.

Data is essential not just for traditional AI models that forecast and analyse, but also for generative AI systems that rely on comprehensive datasets for training and context building. However, the challenges around data are many. In some cases, data is available only for a few sectors, which leads to partial or even misleading insights. In others, data is not being collected at all, either because its value is not fully understood or because the mechanisms to collect it do not exist. Sometimes, data is gathered inconsistently or stored in incompatible formats, making it hard to use. Historical data, which is critical for identifying long-term patterns and predicting future events, is often missing or incomplete. There are also issues of poor data sharing across institutions and jurisdictions, which limit the emergence of collaborative solutions. Perhaps, most worryingly, valuable datasets are sometimes held by private companies and placed behind paywalls. This makes it expensive or even impossible for innovators and researchers to access the information they need to build climate, or for that matter, any tech-based solutions. Such barriers ultimately slow down progress and reduce the effectiveness of AI in addressing climate challenges.

With the pace at which climate change is progressing, unlocking high-quality, accessible and well-integrated data becomes crucial to develop solutions to combat it. Without strong data foundations, even the smartest algorithms will fall short. This is where Digital Public Infrastructure (DPI) comes into play. *Remember this term, we will come back to it soon.*

FOUNDATIONAL PROBLEM (READ: OPPORTUNITY) 3: DATA SOVEREIGNTY

Imagine asking for help in your local language and getting an answer in a dialect you don't speak, with advice that doesn't fit your customs or context. That's what using foreign-trained AI can feel like in many parts of the world. If AI is not trained on local languages, climate patterns or farming techniques, it may sound accurate but fall short of being useful. For instance, offering flood solutions built for European cities to Jakarta does little good. This is why countries must also be able to build and train their own AI models rooted in local realities.

When AI models are built mostly on selective data, they often misread or ignore the unique needs of other regions, whose data was not used to train them. They may suggest irrelevant policies, misunderstand risks, or give advice that does not match local systems and cultures. This leads to weak insights and missed opportunities. In contrast, open AI models can be trained on local data, languages and contexts. The overarching objective here is not just about achieving fairness or independence, it is about making AI useful, effective and meaningful where it matters most.

THE BIG 3 AND THEIR UNIVERSALITY

These three problem statements above are actually transformative opportunities, which require transformative solutions. Over the next three chapters, we will take a deep dive into each of these opportunities through a solutions lens.

The three transformative opportunities for building scalable, sovereign and sustainable AI for climate, therefore, are:

1. **Green Data Centres of the Future** (addresses AI's carbon footprint)

2. **Digital Public Infrastructure for Climate** (addresses data gaps)

3. **Sovereign AI** (contextually aware AI)

Perhaps, one of the most compelling things about these three opportunities is the fact that they apply not just to the AI–Climate nexus, but also to the nexus of AI and anything else, and to AI itself.

Climate, or no climate, data centres will still be required, and they do have an environmental impact that needs to be managed. The need to address data gaps exists across sectors, be it health, education, agriculture or anything else. By making data open across these sectors, we will be unleashing the next wave of transformative innovation. The problem of contextually irrelevant AI insights and closed AI models is equally concerning. *Therefore, we believe that these three opportunities are universal for all things that leverage AI.*

The next three chapters offer a potential blueprint for turning these opportunities from vision into reality.

PART 3
THE BLUEPRINT

13

DATA CENTRES FOR THE FUTURE

Building Green Data Hubs, Empowering Communities

Think about data centres as the brains powering AI. What do they need to run? Just like humans, they need energy and they need to keep cool. They require electricity and a large amount of water, and they need it twenty-four hours a day, seven days a week. But if the large amount of energy they consume is not clean energy, these data centres contribute directly to climate change. This surge is not just driven by the acceleration of internet access, streaming and e-commerce. It is predominantly driven by the explosion of AI workloads, which are far more energy-intensive than traditional, digital and cloud services. The rise of generative AI, with its constant demand for real-time computation, accelerates this trend even further. This scale of energy and water use reveals a paradox.

While AI and cloud computing are often hailed as tools to fight climate change through smart electricity grids, precision agriculture, smart transport systems, and many other solutions, which we have dedicated entire chapters to, their carbon footprint may actively worsen it if left unchecked. Without urgent shifts to clean energy and more efficient data centre design, the very infrastructure powering our digital future could deepen environmental stress.

The challenge is not whether we build data centres. This is not even a question. AI is the current frontier of disruption, and stifling innovation has never been the answer. The real question is how to make these data centres truly green, and ensure that the world continues to decarbonize rather than see the climate worsen as they grow. It is just as important to ensure that the benefits from building these data centres do not stay confined to the digital economy but spill over into the wider economy. With the right policies and planning, the jobs, skills and infrastructure they create can strengthen local communities and allow the people living around these facilities to share in the opportunities that the data centres generate.

In this chapter, we explore what data centres are and understand their significant consumption of energy and water. We highlight remarkable innovations in green data centres around the world. We then propose ideas that connect the growing demand for data centres with broader socioeconomic benefits, potentially creating a win-win proposition. Finally, we suggest enablers to accelerate the development of sustainable, green data hubs.

THE SILENT GIANTS POWERING THE INTERNET

Data centres are the unseen backbone of the digital world. Every time we ask a voice assistant a question, stream a video or generate an image using AI, we are relying on data centres, which are vast spaces filled with servers and computing equipment. In simple terms, data centres are like the 'brains' of AI. They are where all the heavy lifting happens: data is stored, algorithms are trained and digital services are delivered in real-time. Just as the brain needs oxygen and nutrients to function, data centres need power, cooling and GPUs or TPUs (the specialized processors that handle complex AI and computational tasks) to stay online and efficient. The rise of AI has supercharged the demand for data centres. As AI applications become more sophisticated and widespread, right from personalized learning tools to autonomous vehicles, the need for computing power has exploded. As per a report by the Dell'Oro Group, data-centre capital expenditure is projected to more than double from $430 billion in 2024 to $1.1 trillion by 2029.[1] Training large AI models like ChatGPT or image generators consumes immense amounts of energy and computational resources. The result is a booming global demand for data centres. In fact, some projections estimate that data centres could account for up to 8 per cent of power in the US alone by 2030.[2] Much of this expansion in data centres is now targeting the Global South. Countries across Africa, Latin America, South Asia and Southeast Asia are becoming attractive markets for new data centre investments. There are good reasons for this: younger, rapidly growing populations are demanding digital services, governments are expanding connectivity infrastructure, and land and energy costs are often lower than in the Global North. Increasing legislation around data sovereignty is also pushing companies to store and process data within sovereign borders, further accelerating the demand for local data centre capacity. A notable example is the World Bank's investment in the Raxio Group, which is building a network of energy efficient, carrier neutral data centres across Africa, including in Uganda, Ethiopia and Côte d'Ivoire.[3] These projects, besides connecting people, are laying the foundation for local AI innovation and digital economies in Africa. Another major development is Google's announcement of a USD 15 billion investment to build its first large-scale AI hub in India, which will include about 1 GW of data centre capacity in Visakhapatnam.[4] Indeed, data centre infrastructure can strengthen digital ecosystems and support broader economic growth.

The interesting opportunity for the Global South lies in the greenfield infrastructure that is emerging, allowing these regions to skip legacy constraints and seize promising new possibilities. Data centres are now being established in this part of the world, creating a real chance to build them from scratch and design them to benefit the surrounding communities as well. Unlike data centres already built in the West, where retrofitting is often costly, this greenfield aspect is especially promising.

GIANT GUZZLERS

However, this growth comes with significant environmental trade-offs. Data centres consume vast amounts of electricity, often drawn from fossil fuels, and use enormous quantities of water for cooling. To put this in perspective, a massive data centre can use as much water per day as three average-sized hospitals or an Olympic swimming pool. In the United States alone, data centres consumed around 1.7 billion litres of water per day in 2021,[5] a figure expected to rise as demand for computational power increases. In water-scarce regions, this raises important sustainability and equity concerns. As per a number of forecasts, electricity consumption by data centres will more than double between 2022 and 2026, largely due to the energy-intensive nature of AI workloads.[6] Various reports, estimates and analysis also project that electricity consumption by data centres worldwide may roughly equal the annual electricity use of Japan today.[7] In another 5 years or less, they could consume as much power as India. A number of analysts believe that almost 5 per cent of the global energy generated would be used up by these data centres soon.

GREEN DATA CENTRES

Given the significant concerns around data centres drawing enormous amounts of power and guzzling water, the associated impact that this might have on our shared and collective future, and growing voices of concern across the world, there is strong interest among companies to work on reducing, minimizing or eliminating the carbon footprint from these centres altogether. Thus, the push for making data centres sustainable.

But what exactly makes a data centre sustainable or green?

Firstly, what powers the data centre is important. If it is powered by energy from a national grid fed by coal-fired power plants, then it is not a clean or sustainable data centre. It should instead be powered by clean energy such as solar, wind, geothermal or hydropower energy, either directly or through the grid.

Secondly, it has to be energy-efficient. This means that, in an ideal world, it should use only as much electricity as needed to perform its computing tasks. It should avoid wasting energy through inefficient cooling or by using equipment that could perform the same tasks with less energy if replaced by modern, more efficient alternatives. An important measure to evaluate the efficiency of data centres is the Power Usage Effectiveness or PUE. Put simply, this is the ratio of the total energy consumption of a data centre (including IT equipment, cooling systems, etc.) to the energy consumption of just the IT equipment it houses. An ideal PUE score is 1.0, meaning every bit of energy consumed goes directly into computing, rather than some or a lot of it going towards keeping the data centre cool. In practice, the world average PUE is around 1.56.[8] Modern green Data Centres are pushing that number closer to 1.1 or even lower.

Thirdly, their consumption of water needs to be minimized. Servers generate heat continuously and need to stay cool to keep functioning. Traditional cooling systems, which often rely on energy-hungry air conditioners and, in many cases, large amounts of water, are highly resource-intensive. This is driving companies to find innovative ways to cool data centres by using the natural environment, recycled heat and minimal water.

COOLEST DATA CENTRES OF THE WORLD

Coolest, literally and figuratively. These centres are 'cool' because of how they manage temperature with impressive efficiency, but also because they represent some of the most innovative and climate-conscious (read: coolest) designs in today's digital landscape. A disclaimer: this is not a ranking or an exhaustive list, it is a curated set of examples that we chose, as the authors. Our aim is simply to make the world of green data centres a little more accessible, relatable and interesting for our readers.

Meta's data centre in Luleå, Sweden, draws in cool Arctic air as part of a technique known as 'free air cooling'.[9] Rather than relying on air conditioning, the facility simply draws in naturally cold outside air to keep its servers at an ideal temperature.

Thanks to this, it achieves a PUE of just 1.07, making it one of the most efficient data centres in the world.[10]

In 2009, Google purchased an old paper mill in the small Finnish coastal town of Hamina and converted it into a data centre.[11] There, it pumps in cold seawater from the Gulf of Finland to cool down its servers.

An emerging frontier in computing looks beyond the Earth itself and into space! Early experiments are beginning to test the idea of orbital data centres. The Nvidia-backed startup Starcloud has demonstrated that AI workloads can be run in orbit, reporting that it successfully deployed GPUs aboard a satellite and executed and queried an AI model from space.[12] This raises a provocative possibility: could data centres be imminent in orbit? Continuous solar exposure above the atmosphere and the ability to shed heat through radiative cooling in the vacuum of space could, in principle, alter some of the energy constraints that define terrestrial data centres.

Interestingly, Starcloud is not alone in exploring this question. A small but growing set of space and technology companies, including ventures associated with Jeff Bezos, have begun to examine whether elements of future digital infrastructure could one day operate beyond Earth.[13] At the same time, such ideas warrant careful scrutiny. Any serious assessment of orbital data centres must account for their full lifecycle footprint, including the energy, emissions, and material intensity associated with manufacturing and launching hardware into space. The question, therefore, is not whether orbital data centres represent a near-term alternative, but whether, over time, their end-to-end environmental impact could meaningfully compare with that of increasingly constrained, land- and water-intensive data centres on Earth.

Microsoft's experimental Project Natick involved placing an entire data centre on the ocean floor.[14] This allowed the cold seawater to naturally absorb the heat generated by the servers. The project was deployed in phases to study the feasibility of underwater data centres. Microsoft's tagline for this concept is '50 per cent of us live near the coast, why can't our data?'

In the first phase, they showed that it is possible to successfully operate data centre equipment underwater. The prototype addressed challenges like cooling high-density electronics and preventing marine organisms from building up on surfaces. In the second phase, Microsoft demonstrated that full-scale undersea data centre modules could be manufactured cost-effectively and deployed in less than ninety days. After retrieving and analysing the Phase 2 unit, Microsoft confirmed that underwater data centres are viable and can offer significantly higher reliability than comparable land-based facilities. Interestingly, the underwater facility recorded just one-eighth the failure rate of similar land-based centres.[15]

Techniques using outside air for cooling can work in colder climates. The air in Sweden and Finland is naturally cold for most of the year, making free air cooling a practical and sustainable option. Submerging data centres may also take some time to become a scaled reality. However, what works in one country or context may not work in another. For instance, you cannot think about using outside air to directly cool data centres in India, Nigeria or Brazil, where the air is hot and often humid. In such regions, trying to cool servers with ambient air would be both ineffective and inefficient. That is why there is no one-size-fits-all solution when it comes to sustainable data centres. They just need to be customized to be climate-appropriate.

In hotter or drier areas, one increasingly common solution is a method called Direct Evaporative Cooling or DEC. This pretty much works like sweating on a hot day (as uncomfortable as it is for us, it may be good for a data centre!). Warm air is passed over damp surfaces, and as the water evaporates, it cools the air. It is a natural process and does have relatively higher energy efficiency, but it consumes water. To minimize water use, some data centres use a variant called Indirect Evaporative Cooling (IDEC). In this approach, the air that cools the servers never comes into direct contact with water. Instead, it passes through a heat exchanger cooled by evaporation, reducing humidity risks and conserving water.

Amazon Web Services (AWS) plans to use a mix of air and evaporative cooling techniques in its data centre in Chile. According to Amazon, their data centre will use water only about 4 per cent of the year, which is roughly equivalent to the amount used by eight homes over 15 years.[16] This approach is especially important in Chile, where a persistent drought is drying up water supplies.[17] This drought, caused by human factors and policy issues, has been worsened by climate change. In any country facing such a crisis, the government naturally focuses on solving this immediate problem, diverting attention from long-term climate change measures. This reduces preparedness for future climate events, creating a *vicious cycle* that is hard to break. With such complex challenges, we must ensure our data centres do not make things worse.

In developing countries, the future lies in building data centres that are both efficient and adapted to local realities. In South Africa, for instance, Africa Data Centres is building a solar farm to power facilities in Cape Town and Johannesburg with clean, carbon-free energy.[18] This reduces dependence on the national grid and helps lower emissions in a country where power shortages and fossil fuel use are still widespread. In Kenya, the Olkaria Ecocloud Data Centre will be powered entirely by geothermal energy.[19] This clean, reliable energy source is abundant in the region and provides a stable power supply without emissions. In

Côte d'Ivoire, the MainOne data centre in Grand Bassam is part of a broader push to create energy-efficient digital infrastructure across West Africa.[20]

In Malaysia, YTL is building a massive 500MW Green Data Centre Park in Johor, which will be powered by solar energy. Johor is fast emerging as the world's preferred hub for data centres, with billions of dollars flowing in as investments.[21] AirTrunk, another company operating data centres in Malaysia, is combining rooftop solar panels with smart cooling systems to reduce electricity use.[22] These systems take into account Malaysia's hot, humid climate and use a mix of airflow design and evaporative cooling to keep energy consumption low. In the Indian state of Maharashtra, the government has proposed the setting up of the Green Integrated Data Centre Parks project, totalling 1.5GW of IT capacity with 100 per cent of the energy demand being met with renewable energy.[23] In Australia, OVHcloud's data centre in Sydney has adopted 'direct-to-chip' cooling technology.[24] Instead of cooling an entire room or even whole servers, this technique delivers coolant directly to the hottest part of the machine, drastically reducing both water and energy use.

Adding another layer of intelligence, many companies are now using AI to optimize their energy and water use. AI systems monitor servers in real time, predicting where hotspots will form and adjusting airflow or cooling dynamically. Google used its DeepMind AI to reduce cooling energy by almost 40 per cent in some of its data centres.[25] These intelligent systems improve over time, learning how to fine-tune operations and eliminate waste. This is a prime example of how AI can solve challenges it creates by addressing its own environmental footprint.

This leads us to conclude that when it comes to AI and data centres, we must create a positive (or virtuous) cycle instead of a *vicious* one. We need to harness technology to solve the challenges it brings. We are already seeing this with green data centres pioneering unprecedented levels of innovation. From underwater facilities to solar-powered campuses, these data centres are creating new standards for reducing power use, cutting emissions and reimagining sustainable infrastructure for the twenty-first century.

But can we take this positive cycle further? Can these data centres, while solving climate challenges, also bring benefits to the communities around them?

If designed with intention, the answer is yes. This could create a true win-win, where innovation drives both environmental sustainability and local socioeconomic growth, building resilient futures for people and the planet alike.

WHEN EVERYONE WINS

Imagine a small hamlet in a developing country. It has just received internet connectivity through optical fibre cables (and it is not entirely untrue that the internet is reaching many far-flung areas faster than necessities like grid-connected electricity and piped, clean water). It has plenty of sunlight, but it does not yet have basic infrastructure to recycle water that is being discharged from homes, nor is it well-connected to the electricity grid. Because it is far away from major towns, making it a hassle to get there, prevailing land rates are quite low. Now, a major tech company decides to build their data centre there, probably because it is cheap to do so. Considering the hamlet does not have a stable grid connection, the company and the government decide to set up a large solar farm. However, solar panels cannot generate power at night. To tackle this, they build battery farms where large batteries store power generated during the day and release it at night. They also add a pumped-storage system, which uses surplus daytime solar power to pump water uphill and then release it to generate electricity after sunset. Working together, these solutions ensure that the hamlet has reliable round-the-clock clean power. But this energy may not always be enough to power the data centre, so they connect the hamlet with the electricity grid, building transmission towers and drawing wires that bring electricity to homes. During off-peak hours, excess energy from solar panels also powers homes and flows back to the grid, making the overall energy mix greener. The hamlet lays down sewer lines to recycle water. The recycled water keeps the data centre cool, and also provides treated water back to the village, which helps with agriculture and daily chores. Perhaps, the heat from the data centres (yes, they produce a lot of heat) can be channeled into homes to keep them warm during winters through pipelines, a model known as district heating.

CLIMATE CO-BENEFITS

In our hypothetical situation, investment in this green data centre brings many benefits to the community besides climate goals. For instance, reliable power could support local entrepreneurship and ease burdensome tasks, giving people time to upskill and innovate. Treated water leads to better health, enhancing productivity. Waste heat can help schools operate more comfortably in colder months. Stable electricity supports better equipment at medical facilities. Internet connections open doors to learning, e-commerce and remote work. New roads built for the data centre provide vital links, connecting the hamlet to broader economic networks. Therefore, Climate Co-Benefits emerge when investments in climate (in this case green data centres) generate benefits that go beyond climate and help improve economic opportunities, health, education, productivity and overall quality of life.

The government should recognize this as a win-win model and incentivize more data centre projects in similar locations. What can be achieved is remarkable: the company operates a clean-powered data centre in a cost-effective location, while the hamlet gains grid connection, water recycling and solar power.

This is not just a data centre of the future. It is a data centre *for* the future.

DATA CENTRES FOR THE FUTURE

In order to bring about these development gains, it is critical to rethink how we design data centres. The goal should go beyond reducing carbon and water footprint to ensuring that these facilities deliver wider environmental and social benefits, especially for the communities around them.

We already see promising examples worldwide. Amazon Web Services cools some of its data centres with reclaimed wastewater that undergoes a multi-step treatment to remove almost all impurities. The company aims to be net water positive by 2030, and has already reached 53 per cent of that goal.[26] After cooling, the water returns to the treatment facility for reuse. This approach addresses both the water needs of data centres and local waste management challenges. Remember Google's data centre in Finland, which uses seawater for cooling? Well, the city and Google plan to capture the waste heat from the data centre and use it to warm local homes, schools and public buildings, meeting 80 per cent of heating demand.[27] Since the data centre runs almost entirely on carbon-free energy, the heat supplied is also nearly carbon-free. Compared to many cities that rely on gas and electricity for heating, this is a major win. It shows how well-designed data centres, innovation and public partnerships can tackle climate issues and support communities. Imagine the impact of such solutions in the Global South where vulnerable communities lack heating infrastructure.

In the United States, tech companies like Microsoft and Google co-locate data centres with renewable energy sources such as wind and solar farms. Microsoft recently committed billions of dollars to renewable projects that aim to generate over ten gigawatts of power.[28] Google partnered with an energy company to power its Nevada data centres with geothermal energy.[29] These represent massive strides towards ensuring clean power and accelerating the growth of sustainable infrastructure.

These models offer great promise for the Global South. Imagine a midsize town with unreliable electricity and limited infrastructure. Building and co-locating data centres with renewable energy and energy storage solutions could help power local communities, besides feeding back to the grid. Cooling might use recycled water from new sewage treatment plants, providing cleaner water to the community. Along the way, the area gains better energy, water, digital connectivity and jobs. These ripple effects, where climate-positive investments support broader development goals, are climate co-benefits.

Such infrastructure must also resist climate risks. Many Global South countries face extreme heat, flooding and drought. AI can help analyse weather and land data to guide where to build data centres safely and adjust cooling systems ahead of heatwaves.

These insights call for a fresh vision of digital infrastructure. Data centres can become catalysts for green transformation as well as socioeconomic development, both of which are crucial for the Global South.

An ideal data centre is energy-efficient, powered by carbon-free energy, minimizes water use and is intentionally located where green measures undertaken can bring about socioeconomic benefits for the communities around them, creating a win-win situation. It must be climate-resilient, sited where risks from climate-driven extreme weather are lowest and adaptive enough to handle unexpected events.

As AI workloads scale and data-centre demand accelerates, the primary constraint may no longer just be chips, land, or capital, but grid capacity itself. In many regions, transmission networks were never designed to accommodate large, concentrated, and always-on loads of this magnitude, especially alongside rising shares of variable renewable energy. Long interconnection queues, delayed transmission upgrades, and limits on local substations are now slowing data-centre deployment as much as planning or permitting processes. This is forcing an infrastructure rethink, and is pushing developers and hyperscalers to look beyond traditional grid reliance towards co-located generation, storage, and demand-side flexibility. In this emerging landscape, access to reliable power is no longer a background utility consideration, but a central determinant of where and how digital infrastructure can grow.

New power-supply models for data centres are also emerging, reflected in Alphabet's announcement of an USD 4.75 billion[30] agreement to acquire Intersect Power, a developer focused on building large-scale energy infrastructure alongside data-centre demand. Intersect's operating model focuses on augmenting today's constrained grid by co-locating demand with power sources, enabling growth at scale. In the context of Data Centres, such a strategy signals a shift away from treating power procurement as an external constraint and towards planning compute capacity and energy infrastructure together. Bringing energy development capabilities closer to data-centre strategy can help reduce exposure to grid bottlenecks, shorten deployment timelines, and better align the rapid growth of AI computing with the availability of reliable power, while grids themselves catch up.

Many technologies discussed in this book such as geothermal, solar, wind, pumped storage and batteries will play key roles in this shift towards green and community-positive data centres for the future. Countries should carefully plan data centre locations to maximize climate co-benefits and kickstart green development. This requires policy incentives and streamlined regulations to ease challenges like land acquisition, grid integration and clean energy certification. Removing these barriers is essential for sustainable data centre growth worldwide.

CALL TO ACTION

1. **Create a National Green Data Infrastructure Pipeline (DIP):** Identify and map locations for building data centres. Key considerations should include a comprehensive score based on land availability, ease of accessing carbon-free energy, available government incentives and climate co-benefits for local communities. This includes benefits from renewable energy capacity, grid integration, energy storage and wastewater use associated with the data centre. One of the most important considerations should be climate risk: not just current exposure to extreme weather like floods or heat

waves, but projections for the future. This is crucial. We do not want to invest billions in data centres only for them to be at risk from such threats. This pipeline will attract companies interested in investing in data centre infrastructure.

2. **Establish Climate Risk Monitoring Systems:** Risk assessments should not be one-time exercises. With climate change accelerating, monitoring must be continuous and adaptive. Data centres, and the areas where they are located, should be equipped with sensors to track environmental conditions using advanced technology, AI and GIS tools. Data should feed into a centralized platform for real-time climate monitoring and proactive responses. For example, preemptive cooling can be activated during a forecasted heatwave to avoid downtime. Over time, the platform could include other monitoring like cyber risk. This will help keep data centres secure, resilient and operational in an increasingly volatile world.

3. **Enable Co-location of Data Centres with Clean Power:** As grid constraints tighten, countries must expand the range of viable power models for data centres beyond traditional grid-only approaches. Co-locating data centres with renewable energy, storage, and firm supply sources can ease pressure on overstretched networks while accelerating deployment. This includes enabling large-scale renewable energy parks designed around data-centre demand, as well as exploring emerging firm, low-carbon options such as small modular reactors where appropriate regulatory frameworks and public acceptance exist. Recent moves by hyperscalers to acquire or partner with energy infrastructure developers

reflect this shift towards planning compute and power together. Governments should support such integrated models through clear siting guidelines, streamlined approvals, and market rules that recognize data centres not just as consumers of electricity, but as anchor loads around which clean, reliable power systems can be built.

4. **Make Carbon-Free Energy Easier to Procure:** Carbon-free energy is critical for green data centres. Barriers such as limited grid access and the absence of real-time energy tracking still make it difficult to procure this energy at scale.

    The focus should also shift to the procurement of clean power on a real-time basis, versus undertaking an annual matching exercise. We must move from 'accounting green' to 'real green'. Energy storage projects should complement other renewable sources such that 24×7 clean power can be made available to data centres.

5. **Rethink Grid Operations for an AI- and Renewables-Heavy Future:** As outlined earlier in this chapter, grid capacity is fast becoming the binding constraint for both clean energy deployment and data-centre growth. Addressing this challenge requires rethinking how electricity systems are planned and operated. Governments and system operators must accelerate the transition to more intelligent, flexible grids by upgrading transmission infrastructure, reducing interconnection delays, and embedding advanced digital and AI tools into grid operations. AI-enabled forecasting can improve the accuracy of renewable generation estimates, lowering reserve requirements and system costs. Real-time congestion management, smarter battery dispatch, and

predictive maintenance of transmission and distribution assets can reduce outages, losses, and curtailment. These measures can unlock latent grid capacity and ensure that high-growth digital infrastructure and renewable energy can scale without compromising reliability or affordability.

6. **Incentivize Building Green Data Centres:** Companies using cutting-edge technologies to improve energy efficiency and sustainability should receive incentives like tax breaks, carbon credits, subsidized renewable energy rates, green certificates for government procurement, and risk-sharing mechanisms from governments or development banks so that they are able to invest with greater confidence.

7. **Offer Innovative Financial Instruments:** The growth of data centres, particularly in emerging markets, is shaped by high upfront capital requirements and reliance on external power, water and grid infrastructure. Public finance can play a limited, targeted role in reducing early-stage risks associated with shared enabling infrastructure, thereby facilitating private investment. First-loss guarantees can support the adoption of experimental, locally-suited cooling and water-efficiency technologies, while viability gap funding can improve the economics of co-located renewable energy and grid assets. Appropriately structured concessional lending can reduce financing costs, supporting more efficient and resilient private data-centre development.

8. **Create a Single Window Clearance System:** A centralized online portal should allow companies to apply for all permits

and incentives at once, covering land leasing, electricity connections and subsidies. The system should have clear timelines to avoid delays, enabling quick project start. All unnecessary regulatory barriers and clearances should also be identified and rationalized, or eliminated.

9. **Forge Public–Private Partnerships for Infrastructure:** Renewable energy and wastewater treatment facilities should be developed jointly by data centre companies, governments and private firms. Data centres can serve as anchor customers within co-located infrastructure zones, providing long-term, predictable demand for power, water and cooling. This can support investment in renewable generation, storage, grid upgrades and water-reuse systems, while reducing demand risk and speeding up approvals. Shared infrastructure of this kind can also serve surrounding communities, strengthening local resilience and delivering wider public benefits.

10. **Develop a National Data Centre Policy:** All of the components that have been listed out so far should be factored in to the evolution of a National Data Centres Policy. The policy should remain flexible enough to cater to a sector that is fast evolving on the back of disruptive innovation.

THE FINAL WORD

To recap, data centres are no longer just back-end infrastructure. They are the nervous system of the digital world, powering everything from public services to AI innovation. Their growing role comes with a larger carbon footprint. The data centres of the future must be efficient, sustainable and resilient. They should not only consume clean energy but also catalyse green infrastructure for surrounding communities. Climate-aware designs and climate-responsive operations are essential. Governments, tech companies and multilateral institutions all play key roles. By easing investment, encouraging co-benefits, and by leveraging the right policy and finance levers, we can unlock a green and inclusive digital economy.

The next time we share that beautiful image that we generate on ChatGPT, powered by GPUs, it should come from a data centre running on green energy and supporting its local community.

But what about all the data these data centres process to give us answers to questions that we ask ChatGPT or Grok? If we want to build truly impactful and inclusive AI–Climate solutions, we need to take a closer look at the data itself, or rather the lack of it, and ask how we can ensure this foundation is strong and accessible.

14

DPI AND DATA COMMONS FOR CLIMATE AI

India's Digital Inclusion Playbook, Rebooted for Democratizing Climate Data

Before we jump into the data that will heat up our data centres, let us take a quick break from all the talk about climate and AI. If you're feeling a bit overwhelmed already, this change of scene might be just what you need.

UNDERSTANDING PAYMENTS: INDIA VS THE WORLD

For centuries, cash was king: people paid in notes and coins, fumbling with exact denominations and facing theft risk. The search for convenience and security pushed many countries to digitize payments. Sweden became one of the most advanced examples of this shift by expanding mobile and online payments in the 2010s; the Netherlands also embraced electronic payments early on.[1] These developments were followed by broad roll-outs of card payments and mobile wallets in markets such as the United States, China and parts of Europe.[2]

However, even today, in some of the most developed countries of the world, if you want to pay someone digitally you often need a compatible card or a specific app. Merchants may accept only certain payment platforms, peer-to-peer transfers can require bank details or both parties to use the same service, and small vendors without point of sale (POS) machines are frequently excluded. This fragmented model, where payment mechanisms are tied to banks, cards or individual apps, makes everyday transactions cumbersome and limits financial inclusion in many markets.

India had a simple idea: if the problem is incompatible systems, wouldn't it be easier if everything was ... unified?

Welcome to the world of Unified Payments Interface or UPI.

In India, it does not matter which payment app you use or whether you have a card at all. You can transfer money with a single click using only the recipient's UPI ID, which is most often just their mobile number. Merchants may have quick response (QR) codes that you can scan to pay, and users can also generate UPI IDs other than their mobile numbers, which looks something like 'your name@yourbank's name'. But wait, QR codes can be scanned in many countries. How is this model any different? Well, the fundamental difference lies in the fact that it does not really matter which app you use and which bank account you have—everything works with everything else. Forget smart phones, even if you have a feature phone (remember those old and heavy Nokia phones? Of course, they are much lighter now), you can pay by dialing *99# and entering the recipient's mobile number, and it happens instantly, in real time. Payment gets deducted at that moment from your bank account and gets credited to the beneficiary's account.

This is very different from most countries where two people need to have the same app for a payment to be made between them, and you need credit cards to even make accounts on such payment apps and gateways. Interestingly, in many parts of the developed world, cash still remains common. On travels to some Western cities, we found small businesses like barbers or kebab stands that only accepted cash, even in central locations. In contrast, in India, you might find a lone tea stall on a highway that would accept only digital payments. Today, shopkeepers in India look visibly surprised if someone tries to pay them in cash. Such has been the digital payments revolution in India.

THE BACKBONE: DIGITAL PUBLIC INFRASTRUCTURE (DPI)

In India, billions of bank accounts, mobile numbers, and unique identities all interconnected to form a digital highway. Every Indian citizen has a biometric identity card called Aadhar. It links fingerprints, iris scans and photos in a secure database. Aadhar is connected to bank accounts and mobile numbers, forming a trinity of secure and verified identity. Aadhar creates and verifies identity, UPI helps make payments digitally between bank accounts of users, and all of this is based on the consent of users. Powering this is the India Stack, which is the world's largest integrated suite of open Application Programming Interface or APIs, based on the core building blocks of digital identity, payments, data and consent at population scale. How this system transformed digital inclusion in a country of 1.4 billion people is remarkable, and its principles have now come to be known as Digital Public Infrastructure or DPI, and are guiding digital transformation globally.[3]

Too much jargon? Let us make this simpler. Imagine you are a citizen in India who has a verified digital identity, a bank account to enable digital payments (UPI), and a mobile number. Think of this interplay for each user as a trinity. Think of each of these trinities as being billions of points on a digital highway. Every payment app in India builds itself on top of this highway, and that is what makes them interoperable. That is why someone using Gpay can transfer money to someone using PhonePe (a local Indian app), and someone using PhonePe can transfer money to someone using PayTM (another local Indian app). It is because the underlying foundation is the same. It's not only person-to-person or merchant payments. Most online payment gateways also support UPI: just enter the recipient's UPI ID and a payment prompt will appear in the payer's UPI app asking them to accept or reject the specified amount. Forget typing your card details. Many of us in India can't remember the last time we did that. Moreover, even creating accounts on these apps does not require any heavy lifting at all. When you download an app, all you need to do is to enter your mobile number, and since your identity is verified to begin with, it will just double-check by sending a one-time password (OTP) on your phone, and your bank account automatically gets linked. You don't even need to enter your name.

Let us make this easier. Imagine a scenario where every company offering rides between town A and B ended up building their own roads, ran their own buses, charged their own fee and if you had to get from point A to B, you had to use one of these roads end to end. This excluded smaller bus companies from operating on these highways, allowed bigger companies to build monopolies, charge desired fares, and smaller players would never even think about competing in this business model. Passengers would be left with no choice but to choose one of these expensive bus services. Those who were too poor would not bother travelling. This is so counterintuitive that such a transportation model does not exist in the world. Sadly though, that is broadly how the payments ecosystem works across the world today. Private players run it, they have their own systems, they own the data of their customers, and they are mostly wary of interoperability. That is why two people need to have the same app to transfer money to each other. But that is not the case in India, and it is not the case in many other developing countries which are embracing DPI.

DPI means that instead of private companies building their own exclusive roads, the government builds efficient highways open to all. Passengers gather at bus stations along these highways and wait for buses driven by any operator, regardless of size or wealth. Smaller players who could never afford to build their own roads can now run buses here. The many buses compete for passengers by offering lower fares or attractive deals. Passengers benefit from choice and affordability, including those who previously could not afford to travel. Smaller bus companies find opportunity, larger companies innovate and grow, and the government solves multiple problems at once.

Similarly, by building your app on top of this DPI highway, you gain instant access to the entire ecosystem without the cost of creating your own infrastructure. For example, Google Pay in India commands around a 36 per cent market share by leveraging this digital highway.[4] Instead of investing heavily to build payment infrastructure, apps compete by offering innovative services, offers and superior customer experience. The result is healthy competition and increased user choice, all powered by an open, public built DPI.

DPI FUELS INNOVATION

DPI also acts as a catalyst to fuel innovation. With the building of such robust digital highways where users have a digital identity and are financially included, further innovations are unlocked. This is because now we have a base, and that base is strong, robust and authenticated. This is why so many payments apps were built on top of this system, including global players like GPay, which operates on this base layer, unlike in any other country in the world, where people have to configure their cards to use it. Here, it is just linked to your bank account, digital identity and phone number.

With the payments and identity ecosystem now established, many startups began seeing this as an opportunity to offer even more services digitally and seamlessly. Startups offering flow-based lending, personal finance management, investment apps and online insurance services all started booming in a big way. For them, they already had a set foundation. They did not need complex paperwork, bank account details, or anything else for that matter. Users can just sign up, similar to the payment apps, with their phone numbers. With a simple consent-based authentication, their bank accounts and identity automatically get linked to these apps. This created a win-win situation—startups did not have to create their own complicated highways, they just decided to jump on the public highway. For users, instead of complexities, everything now happens at their fingertips.

This integration means financial services avoid duplicating infrastructure and focus on delivering user-friendly affordable products that reach millions, or billions, in the case of India.

This foundation has supported the emergence of numerous startups innovating in these areas. They leverage India's DPI platform focusing on accessible, secure and affordable financial solutions that empower millions while fostering healthy competition and continuous innovation.

In a way, the idea of DPI led to the emergence of a 360-degree ecosystem, where citizens have not only become more financially included, but innovation has also received a fillip as a consequence. It was this cycle of innovation that led to the financial inclusion story to evolve from cash-based payments to cashless payments, and from cashless payments to digital lending and tailored financial offerings. The cycle of innovation continues—and will continue—as newer use-cases are discovered, and newer solutions are built, not just for financial inclusion, but across many other sectors as well.

DPI BEYOND FINANCIAL INCLUSION

In healthcare, for instance, DPI could enable interoperable digital health systems that securely connect patient records across hospitals, clinics and pharmacies. This means healthcare providers get timely access to accurate patient information, which helps reduce errors and avoids repeating tests. With DPI, telemedicine can become more effective, and public health monitoring can improve, making healthcare more accessible and better for everyone.

Similarly, in the education space, DPI can help make certifications and managing academic credentials much easier, more secure and genuine, especially in an age where online education is booming, as is the EdTech ecosystem. Besides these, DPI can also be extended to agriculture, helping connect farmers directly with the market and allowing them access to credit, insurance, and savings products.

Overall, DPI acts like the backbone of a digital ecosystem that drives better governance, transparency and access to inclusive and efficient essential services. It opens doors to innovation, establishes a level-playing field, and creates a win-win situation, where customers, sellers or service providers and governments, all win. It can drive equitable social and economic benefits for all citizens, while putting countries on a high growth trajectory.

GLOBAL DPI MOMENTUM

This remarkable potential of DPI has not gone unnoticed. Recognizing its transformative power, international organizations including the United Nations Development Programme (UNDP), the World Bank, the International Telecommunication Union (ITU) and the United Nations Science and Technology Envoy's Office are actively advocating for and supporting the rollout of DPI worldwide.

Countries from Ethiopia to Sri Lanka, Bangladesh, Brazil, Ukraine and Zambia are adopting and adapting DPI models to drive digital inclusion and socioeconomic growth. For instance, Ethiopia has implemented the Fayda system, a digital identity platform that facilitates service delivery and financial inclusion for millions.[5] Brazil combines its Gov.br digital ID system with the Pix instant payment platform to streamline access to government services and financial support, while Ukraine's Diia app and Trembita data exchange offer integrated e-government services with high adoption.[6,7] Zambia is focusing on interoperable digital IDs and government payment platforms to enhance social protection programmes and public service access.[8]

Inspired by India's comprehensive DPI framework, several other countries are actively implementing similar systems to improve governance and financial access. The Indian model's scalability, openness and secure identity verification through the trinity are often cited as benchmarks. International initiatives now focus on extending DPI's reach to accelerate socioeconomic development globally.

EXTENDING DPI TO CLIMATE AI

With global momentum building around DPI, expanding its scope to include climate data and AI can significantly accelerate innovation. This expansion not only fosters solutions that are inclusive, accessible and cost-effective, but also ensures that the benefits of the AI-Climate nexus emerge more rapidly, and reach all segments of society.

To expand DPI to Climate AI (having established the nexus of Climate and AI by this point in the book, we may keep referring to it interchangeably as 'Climate AI' from this point), we propose the following two routes:

1. **DPI-powered Climate and Energy Stack**

Similar to how DPI powers financial inclusion, agriculture and healthcare, imagine building a Climate and Energy Stack grounded in DPI principles. The use cases are endless.

Imagine neighbours who have solar panels not only producing their own clean energy but also trading the extra power with people nearby. Think of this as an extended use case of the financial inclusion model that we explained earlier. This would lead to the emergence of a local energy market. This, in turn, helps improve efficiency and lets individuals actively participate in the clean energy transition.

At a larger scale, DPI can support transparent and secure markets for carbon credits where countries, businesses and even citizens can buy and sell verified emissions reduction credits. Trust and traceability are critical here. As we discussed in earlier chapters of AI use cases for climate, AI can help verify data, prevent fraud and ensure the credibility of these markets.

DPI can also play a vital role in adaptation. For example, governments could use DPI to deliver conditional cash transfers on extremely hot days to help people cope with the heat. Similarly, payments could be made quickly to farmers whose crops are damaged by floods or droughts. This would rely on verified environmental data to confirm losses and ensure aid reaches those in need when it matters most. Moreover, DPI can enable insurance products linked to verified identities. AI can automate claims processing by using satellite images and secure records, speeding up payments and helping vulnerable communities build resilience.

With different layers like payments, identity verification, user consent and climate and energy data all running on a common digital platform, opportunities for innovation flourish. This enhances inclusion, improves the efficiency of Climate AI solutions, and builds trust because everything is connected and transparent. Without these layers working together, many innovations would remain abstract ideas, and many citizens would be excluded from the benefits of the transition to a greener future.

2. Data Commons

The second proposed extension to DPI for Climate AI is around opening up and democratizing data. As we have seen in countless AI use cases throughout this book, the effects of climate change touch every sector, and no data can be dismissed when it comes to managing its impacts. Weather records, glacial melt data, river flood information, and satellite imagery can be combined to help predict extreme weather events with the help of AI. When this type of data is available, city planners can act in advance, organizing evacuation and emergency response efforts more effectively. The electricity grid can adjust for surges or drops in demand. When digital twins of infrastructure like bridges and railways are matched with this data, the city can decide which ones might be at risk and should be temporarily closed to keep people safe. Maybe a solar farm is likely to go under water, and the grid needs to take that fall in generation into account.

It is not just about the data itself. Insights created by one solution can make another solution stronger, delivering assessments that are timely and relevant. This is how a true ecosystem of Climate AI solutions emerges, where every solution is linked to other solutions, and they work in unison, drawing on insights and data from each other. This is what was missing in Climaville when flash floods hit. You might have noticed that even with all the mitigation steps taken by Mayor Ashman, Climaville still faced the devastating impacts of flash floods. Recovery and rebuilding did take place, but the reality is that the climate is a 'Global Commons', which means it is a shared phenomenon that will affect everyone, and that no one can manage alone. For global challenges like this, what we need is a 'data commons', where data and insights are shared widely so that adaptation can be effective everywhere.

Today, Climaville stands out as a model where data, which can help the city deal with climate, both mitigation and adaptation, is democratized, open, and available for everyone to use. Even the insights from all solutions have been put onto a common portal.

Think about data on air quality, satellite imagery, data from sensors installed on key infrastructure and in geologically sensitive areas, live data on river water levels, temperature data, and real-time data on crop health from hundreds of agricultural fields. Add to this data related to real-time demand on the electricity grid, data on traffic patterns, information on greenhouse gas emissions, urban heat islands, coastal erosion, rainfall, wind speeds, as well as historical weather and disaster data, land-use records, biodiversity indicators, and predictive modelling outputs. There could be thousands or even millions of such categories of data, and all of this matters for Climate AI. Sadly, in the real world, not all of this data exists. Sometimes, it is simply not collected or, if it is, it is not shared, or is stored in formats that are not usable by those wanting to build solutions. In many cases, where large private players collect such data, they keep it behind paywalls, making it inaccessible for many who might benefit most.

This is exactly what is missing across the world today, and this is why we propose that the concept of DPI should be extended to include the principle of Data Commons.

WHAT WOULD THIS ENTAIL?

Simply put, every piece of available data that is non-personal and does not pose a privacy risk, no matter who owns or collects it, should ideally be made public if it impacts climate outcomes.

Governments should clearly define which categories of data qualify as Data Commons, and all such data should be made accessible in the public domain regardless of its source. Over time, this could develop into a real-time data repository or an extensive data exchange that feeds into a comprehensive portal of portals. This platform would unite every category of real-time data essential for climate action. There should be no gatekeepers to this data. It should be the digital highway, similar to what UPI is in finance, enabling innovative Climate AI solutions to be built on top. This would create a powerful foundation for innovation, unlock new ideas for mitigation and adaptation, and empower researchers and communities.

CALL TO ACTION

Achieving what we have proposed is not easy, but it is not impossible either. While there are a number of complexities that are involved, we propose a few foundational ideas that could at least get the ball rolling.

1. **Expand definition of DPI to include Data Commons:** Democratizing data through a Data Commons approach will establish an ecosystem-wide backbone where world-class Climate AI innovations can be developed by all, and for all. Given that the definition and scope of DPI are still evolving, and since DPI is gaining significant global momentum, embracing Data Commons as a core element would be a major leap forward. This would position DPI as a champion of open and democratized data, creating the foundation for more inclusive, transparent and impactful climate action worldwide.

2. **Create National DPI Missions:** Given the multi-sectoral applicability and benefits that can be unlocked through DPI, national DPI missions or task forces should be established. These should bring together resources and expertise from different levels and departments of government, the private sector, research institutions, academia, startups, and other relevant stakeholders. These missions should be responsible for defining what categories of data should be shared under the Data Commons framework. They should establish standards for such data, determine who is responsible for managing it, guarantee interoperability across systems, and lead capacity-building efforts to support effective implementation.

3. **Create National Data Commons Portal, and Global Portal of Portals:** Expansion of the definition and building a clear understanding is just the first step. It also needs to be put into action. Once the definitions, standards, and responsibilities are in place, countries should work to create national portals that gather data from across the entire country. This includes everything from sensors monitoring river flows to real-time information on power generated by solar plants in the remotest locations. All stakeholders, whether government or private, should share data that affects climate outcomes on this portal. But as we talked about earlier, climate is a 'global commons' and its effects do not stop at borders. Therefore, these national portals should feed into a global portal to strengthen collaboration and develop solutions that can make an impact worldwide. Just because your country is not polluting does not mean that ice caps in your region will not melt because of pollution elsewhere. That is why we need solutions that take into account both local and global realities. Of course, it would be important to appropriately manage any climate data points that could be seen as a national security risk, but this should not be a reason to block the overall exercise.

4. **Win Private Sector Trust:** Unfortunately, even today, many in the private sector do not fully understand the concept behind DPI. They become afraid when they hear phrases like 'democratized data' or 'open data'. They immediately assume it is tied only to social good and that they will not be able to make a profit. What they do not realize is that with a ready-made highway in place, they can focus on running bus services and improving their offerings, instead of building the highway themselves. This fear is even stronger in areas like

climate and energy, where many private players do not see an obvious revenue model. Yet the reality is that DPI-enabled climate and energy platforms can adopt freemium models, where basic access is free but advanced analytics, premium tools, or sector-specific services are chargeable. A global-level outreach effort is needed, especially since there are too many different definitions of DPI circulating around the world.

5. **Establish a Global DPI Body:** Because definitions of DPI are scattered widely, and many want to join the DPI movement, which is a positive sign, it also means there is a need for a central convener to bring everyone together. Around the world, every international organization and multilateral agency is seeking to champion DPI. This enthusiasm reflects the undeniable power of DPI, but it also risks fragmenting efforts and losing focus on a unified and harmonized approach. In the realm of Climate, it is essential that solutions developed by different countries can communicate and work together. That is where harmonization plays a crucial role. Therefore, it is vital that these efforts are aligned and coordinated, making the presence of a global convening agency a necessity.

6. **Incentivize DPI for Climate AI:** Countries must pursue twin priorities in building DPI for Climate AI. Firstly, they should actively incentivize the private sector and startups to develop Climate AI solutions on top of DPI while contributing valuable data and insights back into the system. These incentives could include privileged access to select non-personal, non-confidential government datasets

that can accelerate innovation, participation in regulatory sandboxes that allow faster experimentation, and preferential onboarding onto national platforms or marketplaces. Financial incentives can play a supporting role as well, although they are rarely sustainable in the long run. Secondly, countries must collaborate with private entities and startups to design and build the foundational framework of such DPI. This involves creating the portal of portals, ensuring interoperability across systems, and defining seamless API integration. Both priorities are essential and mutually reinforcing, as innovation and collaboration can only thrive on a well-designed, inclusive digital highway that engages all stakeholders from the outset.

7. **Reciprocity and Data Commons:** At the risk of sounding controversial, we still believe this needs to be said. Climate change is a global problem, and every small piece of data counts. Governments cannot be expected to produce all this data alone. Consider the air purifiers in people's homes that record real-time air quality, the temperatures recorded by air conditioners, and traffic patterns on the roads. All of this data is vital for Climate AI solutions. Unfortunately, much of this data is owned by private companies, creating separate digital highways once again. Even if incentives fail to convince private companies to share such public-good data on a common platform, we could explore other alternatives. Companies know the value of the data they would receive through DPI because it brings thousands of data points at no cost. However, if they want to use and build on this data, their access should be conditional on their willingness to contribute valuable data points back, which are crucial for addressing climate change.

8. **Build National Climate and Energy Stacks:** Countries should create a Climate and Energy Stack based on DPI principles. This could, for instance, aid emergence of local energy markets where individual users and businesses can trade power and carbon credits, the government could make financial transfers during climate-triggered events, and getting insurance for climate events such as extreme heat becomes easy for the citizens. In order to achieve this, governments must develop digital identity solutions for climate-affected vulnerable populations and also promote DPI literacy.

9. **Mainstream DPI into National Climate Plans:** DPI should not be viewed merely as a technical or IT priority. Its greatest impact comes when it is applied across multiple sectors. Climate change affects every sector, and to combat it effectively, all sectors must act together because of their interdependencies. Climate AI solutions need to be integrated into national climate plans. At the core of the success of Climate AI is DPI. Therefore, DPI must become a fundamental component of national and global climate plans to drive effective climate action.

10. **Leverage AI for DPI:** So far, we have emphasized using DPI as the foundation on which Climate AI can thrive. However, AI itself can also play an important role in building and strengthening DPI. AI can help automate and optimize many aspects of DPI development and management. For example, AI-powered data pipelines can ingest, clean and organize massive volumes of climate and environmental data in real time, ensuring high data quality and consistency. Machine learning algorithms can analyse diverse datasets to

detect patterns, fill information gaps, and provide predictive insights supporting better decision-making. AI can also enable intelligent interoperability by standardizing data formats and facilitating seamless API integrations between heterogeneous platforms.

THE ROAD (READ: HIGHWAY) AHEAD

As we move forward, it is useful to reflect on the key takeaways from this chapter. DPI and its extension into Data Commons are vital engines for accelerating climate action. With a well-designed, interoperable infrastructure in place, we will set the stage for the rapid development of impactful Climate AI solutions that address complex environmental challenges. This will create a common digital highway on which locally relevant Climate AI solutions can be developed and scaled globally.

A crucial caveat: none of the data sharing discussed here involves personal data. Data such as air quality, temperature and traffic flows are not personal information. While some data may be linked to individual uses, such as navigation apps tied to specific users, the data we refer to should be anonymized and aggregated to extract relevant environmental metrics without any personal identifiers. Today's technologies ensure privacy is safeguarded while enabling impactful data sharing. So, there is no cause for concern about personal privacy being compromised.

We also stress that the approach recommended here can go a long way in ensuring a well-governed, inclusive and scalable DPI ecosystem that serves the public good across a wide range of applications, not only Climate AI despite the focus of this book.

Finally, with these foundations in place, we must confront one more critical issue, which is also the next step forward. Should an AI model trained on data from the United States offer climate insights to Zambia or India? This raises the urgent question of sovereign AI models. Data sovereignty does not mean closing off data or resorting to protectionism. Instead, it means ensuring contextually relevant data guides your solutions. How can we ensure that Climate AI models provide insights that are truly relevant and accurate for each setting? Because, quite literally, the right insights can mean the difference between success and failure, resilience and disaster, and survival and ruin.

15

DECOLONIZING AI

Championing Sustainable Development Through Open and Sovereign AI Solutions

In early 2025, the AI landscape was shaken when China-based DeepSeek released DeepSeek-R1, powerful large language model that rivalled some of the world's best at a fraction of the cost.[1] What truly captured attention was DeepSeek's decision to open-source these models, sending a strong message that frontier AI need not be proprietary, expensive or secretive. This combination of competitive performance and employing openly published architectures sent ripples through Silicon Valley and policy circles alike.

DeepSeek's success came from embracing openness: building on pre-trained models, leveraging publicly available data, and employing transparent architectures. This approach slashed costs and sped development.[2] While companies like OpenAI and Google reportedly invest hundreds of millions to train models like GPT-4 or Gemini, DeepSeek achieved competitive results far more cheaply. This was possible because rather than reinventing every component, it built atop open resources shared by the community and contributed back, creating a virtuous cycle of reuse and improvement that stands in contrast to the typical closed model race. Today, as per the Harvard Business School, the open-source ecosystem's supply-side value is estimated at 4.15 billion US dollars with demand-side impact reaching 8.8 trillion US dollars.[3]

UNPACKING OPEN VS CLOSED SYSTEMS

Before we proceed further, it is important to unpack a bit of this jargon. (If you already know this, or work on AI LLM models, our apologies in advance for a very rudimentary explanation that is to follow.) So, what exactly is an AI model?

An AI model, particularly a large language model, or LLM, is essentially a mammoth statistical engine of sorts. You can train it by exposing it to tons and tons of data: text books, websites, academic papers, code, even manuscripts. Through training, it learns how language works, how ideas are connected, and how logic flows. Training a large language model is like teaching a student by giving them access to every book in the world, every newspaper ever printed, and every public talk available online, all at the same time. The more they read, the better they get at understanding, summarizing and generating responses to questions across topics. Of course, this student would need to have infinite energy, infinite time and an infinite concentration span (we definitely were not like this at school). GPT-4, for instance, was estimated to cost upwards of $100 million just in computing resources. This figure is reflective of the costs associated with the hardware—thousands of high-performance graphics processing units (GPUs) or specialized AI chips—that had to run continuously for weeks or even months to train the model. These chips performed trillions of calculations on vast datasets, consuming enormous amounts of electricity and requiring sophisticated infrastructure to keep them operational (no wonder data centres are gobbling up energy at this rate). In simple terms, training such a model demands a supercomputer-scale setup, and the cost of renting or maintaining that computing power adds up quickly. In contrast, open models often cost less because they build on prior work, reuse architectures, and benefit from community-driven innovation.

What makes open models different is that they are built in a transparent way. This means, in practice, that their design, how they were trained, and often even the data they learned from are published publicly. This also means that anyone, from a solo developer in Nairobi to a university team in Jakarta, can take these models, adapt them to suit local needs, and put them to use in the real world. Now, this is quite in contrast to how models like ChatGPT or Gemini work. They are usually available only through APIs and monthly subscriptions. While companies sometimes publish papers about these systems, the full model weights and training data are generally not available for inspection or modification without permission or payment, and so they operate largely as 'black boxes'.

At this point, it is important to differentiate open data from open AI. Open data, which is what we discussed in the previous chapter, refers to datasets freely accessible for use. This could be government data, public data, weather data, data from hundreds of sensors, satellite data and so on. Open AI, on the other hand, relates to AI models and software whose architectures, source code, and sometimes trained weights are made publicly available for use, scrutiny and modification. Trained weights are the model's learned parameters that shape how inputs are turned into outputs. While open AI models are often trained on openly available data, not all AI built on open data is open AI, and not all open AI models train solely on open data. The distinction matters because democratizing data fuels AI innovation, but open AI emphasizes transparency and collaboration in AI model development itself.

While the evolution of DeepSeek signals China's emerging AI prowess, we should also see it as the emergence of a model that allows users—companies, citizens and countries—to leapfrog with public goods. This matters enormously for countries in the Global South.

THE CASE FOR OPEN AND SOVEREIGN AI ECOSYSTEMS

Open models also allow nations to retain technological sovereignty. They empower local developers to build tools in indigenous languages, aligned with local contexts, opportunities and challenges. A closed model trained largely on English internet data may often struggle to understand the nuance of Marathi agricultural terms or Swahili metaphors. An open model can be fine-tuned to do just that, and do it quickly, cheaply and without needing permission. However, the challenges are not only on the linguistic front. Closed models may misinterpret local contexts and nuances, they may not understand local governance structures and they sometimes may fail to align with region-specific regulations (like energy subsidies or land rights). It is also quite likely that they lack awareness of climate vulnerabilities unique to a region, such as the water stress in Rajasthan or glacial melt patterns in the Andes. Moreover, models trained on datasets skewed towards Western norms may struggle with cultural appropriateness, local risk perception and nuanced solutions that are deeply specific to complex regional contexts. Real-world evidence shows how AI systems trained largely on Western data often fail to understand Global South contexts. A University of Michigan analysis found that leading image–text models perform much better on scenes from wealthy Western environments than on images from lower-income or non-Western settings.[4] Field reports offer illustrations from the on-ground research of Nigerian scholar, Jake Okechukwu Effoduh, who travelled across Kenya and Nigeria to study how

AI systems behave in real-world Global South contexts. Based on qualitative, interview-based observations from his fieldwork, Jake found that some deployed AI tools misclassified local phenomena. For instance, in Kenya, dairy farmers using AI tools to assess livestock health found that the system repeatedly mis-labelled local cattle as undernourished because it had been calibrated to Western breeds, which are larger and differently proportioned. In Nigeria, Effoduh observed bureau-de-change agents testing an AI application for foreign-exchange monitoring, only to see the Nigerian naira classified simply as 'other', despite the software recognizing dozens of widely traded Western currencies.[5] A study by researchers from Cornell University, KTH Royal Institute of Technology in Sweden and the University of Pennsylvania, provides one of the most comprehensive examinations of cultural bias in large language models to date.[6] By comparing model outputs with nationally representative survey data across more than 100 countries, they found that many leading AI systems systematically align with cultural values common in English-speaking and Protestant European societies.[7]

Closed AI models that are trained predominantly on generic or Global North-centric data often falter in the Global South, where their inherent biases and blind spots may potentially lead to serious misalignments. Imagine a climate policymaker in Peru asking such a model about future water stress and receiving generic drought mitigation strategies, completely missing the country's critical need for glacier runoff monitoring due to accelerating glacial melt in the Andes. Or consider a city planner in Jakarta relying on AI-powered flood prediction tools that fail to account for the city's unique combination of subsidence, coastal surge and poor drainage, offering solutions suited to temperate floodplains instead. Many of us have experienced something similar firsthand, even when using tools like ChatGPT or Perplexity. We ask a question tied to a specific project or local context, only to be met with vague, generalized answers that feel disconnected.

AI models analysing satellite imagery may also misinterpret dense informal housing clusters in cities like Lagos or Dhaka as uninhabited or industrial zones, leading to flawed urban planning or missed service delivery. Beyond climate, these models may misread governance structures, potentially offering US-style farm subsidy application advice to a Nigerian farmer, or suggesting maternal health diets built around dairy and cold-climate foods to a frontline worker in rural West Africa.

A recent study examining a maternal health AI app used in Zambia found that well-intentioned technology can inadvertently reproduce existing data biases.[8] Because locally representative datasets were scarce, the app relied on publicly available data sources that lacked the cultural and contextual specificity needed for Zambian communities. As the study notes, AI systems can "inadvertently replicate biases present in the data they rely on," meaning that important practices, knowledge systems and community-based maternal health norms were at risk of being overlooked. The researchers emphasized that for AI tools to genuinely improve healthcare in underserved settings, they must "reflect local realities" rather than depend on generic datasets that narrow the scope of care. The study emphasized that without incorporating diverse, locally grounded data and knowledge systems, AI risks marginalizing the very populations it aims to support. Similarly, in India, the growing use of AI for weather forecasting can be particularly challenging for the Himalayan region.[9] For instance, the region has over 9,500 glaciers of which detailed glaciological studies have only been performed on only about fifty of these as per researchers from the University of Kashmir. AI trained on western data, even if coupled with limited contextual data, can produce incorrect insights which, in turn, will lead to emergence of unreliable early warning systems. The developing world suffers particularly from the lack of contextually relevant local data and must demonstrate greater resolve to address this challenge.

The problem is not just limited to such models and how they perform in and for the Global South countries, but rather, how they perform against people from particular backgrounds. Indeed, the issue of AI relying on predominantly Global North data means that there have been occasions where it has thrown up biases related to colour, race, gender and countries that people come from, sometimes producing deeply problematic and prejudiced outputs.

DRIVING EQUITY AND ACCESS, THE DPI WAY

These examples hint towards a fundamental issue with closed AI models. They can sometimes provide advice that is suboptimal, incorrect or even counterproductive because they lack the contextual understanding and specificity needed. To be truly effective, AI systems must be trained, adapted, and rigorously tested against the complex realities and lived experiences of the regions they serve. Open-source frameworks empower countries and communities to embed their own scientific, social, linguistic, cultural and regulatory contexts into the intelligence itself. The Global South does not need models trained on non-contextual datasets, nor should it have to pay exorbitant amounts for insights that are either ill-suited for their context or repackaged data from their own countries offered by third parties.

It is also worth noting that most data processing still takes place in the Western world, because that is where the bulk of global computing power is concentrated. Most users, however, are in the developing world. Major AI companies are now partnering with telecom operators in these countries and offering their tools for free. But are they truly free? Enormous volumes of user data flow to these companies at almost no cost, and that data is used to train and refine their models. Once the models mature, they will be sold back to the very billions of people who supplied the data in the first place. In effect, data is being extracted from billions at zero cost and will eventually return to those same billions as a product that generates billions in revenue for the companies.

India's unique model of DPI shows us how digital systems can be open source, support easy connection through open APIs, and grow to serve people everywhere. Open AI models could work in a similar way by acting as a basic platform for many areas like finance, education, governance, health and climate. This platform would be free and open for anyone to use, improve and customize to meet the needs of different communities. Other AI systems could be built on top of these open AI models, adding their own local knowledge and improvements, while also sharing their progress back to the main platform. This way, everyone benefits from a shared, growing digital resource—a common core of sorts.

PROFITING FROM OPEN AI, THE FREEMIUM WAY

A key question you may be asking is this: why would someone make their core open-source? If it becomes available to everyone, how will developers make their money? This was touched upon in the previous chapter and is a common concern among companies, though this narrative deserves changing. We explained India's payments DPI in the previous chapter. It builds a digital highway. But how do those who build payment apps on the highway make money? Payments apps carry advertisements and offers that direct people to merchant sites. They also provide merchant analytics, loans and insurance on top of the basic payment services. These value-added offerings sustain and expand the ecosystem while promoting digital payments across the country. Once the highway exists, it is really about how lucrative the bus service can be made to ensure more people hop onto it rather than competitors' buses. By wooing more customers, companies make more money.

The same principle applies to monetizing open-source AI models: the real value lies not just in the models themselves but in the entire ecosystem around them. Think about Linux, the free operating system that anyone can download and use. Companies like Red Hat make money by helping businesses use Linux better.[10] They simply sell support, training and special tools that make Linux deployment tailored to organizations' needs.

Open AI models often perform best on specific hardware or cloud platforms, which creates platform dependencies and indirect monetization opportunities. Many providers adopt a freemium strategy: they offer basic models at no cost but charge for premium features such as stronger security, faster response times, or specialized tuning. Firms also earn revenue by optimizing hardware or designing chips to run large models more efficiently. Some monetize aggregated or anonymized usage data to improve services and build new features, while others sell access via usage-based APIs, subscriptions or enterprise contracts. The open-source community strengthens these platforms: developers build extensions, plugins and tooling that increase engagement and make a platform harder to replace. As users and developers cluster around a platform, demand for paid enterprise services, premium capabilities and cloud hosting grows, driving revenue.

Today, even models that appear free are strategic tools companies use to build influence, secure market presence, and create long-term dependencies.

In summary, AI monetization's future is increasingly freemium: open and free for innovators aiming to create positive change, with paid tiers supporting enterprise users needing robust, scalable, and customized AI services. This balance encourages broad innovation while driving profits and business growth.

THE CLIMATE LINK

Given the urgency surrounding climate change, and the need for data that can help create solutions that are truly inclusive and impactful, both open data—which we covered in the previous chapter—and open AI models hold the key.

Open AI models represent the 'highway' of data and prediction, while Climate AI companies and startups run the 'bus services' that bring benefits to people. A powerful real-world example is 'Prithvi', the open-source AI model developed by NASA and IBM.[11] This model uses nearly forty years of Earth observation data to provide detailed, high-resolution weather and climate forecasts. It can be tailored for specific regions to predict severe weather, climate shifts, or seasonal changes that help governments and businesses plan and respond better. Because it is open and flexible, this model supports researchers and organizations worldwide and can quickly adapt to different climates and needs.

In combination with open data, open AI models can unlock the full power of climate information, making it accessible and useful everywhere. This democratizes climate action, ensuring no nation or community is left behind as we fight this urgent global challenge together.

CALL TO ACTION

Given the critical importance of open AI models not only to advance Climate AI but also to foster innovation that is contextually relevant, impactful, inclusive and respects sovereignty, we propose the following actions. These ideas aim to create an ecosystem where open AI models serve as foundational tools enabling diverse communities and innovators to develop solutions tailored to local needs while promoting global progress.

1. **Treat Open AI as DPI:** Building on our recommendations from the previous chapter, where we asked for DPI to also include Open Data, the definition and real-world implementation of DPI should also be expanded to include open AI models. Countries must build AI solutions on top of DPI rather than rely on foreign black box systems that offer little transparency or control.

2. **Offer Compute and Host Open AI Models:** If open AI models are treated as DPI, they essentially become public goods that bring wide benefits to innovators, researchers and developers. Governments or international agencies can host open AI models on public cloud infrastructure and make them accessible so that startups and small technology companies can use them without investing heavily in their own compute. This must be complemented by providing substantial compute resources such as GPUs and specialized AI processors. Democratizing access to compute, large datasets and AI infrastructure ensures that these capabilities do not concentrate with a few countries, citizens or companies.

3. **Sandboxes for Safe Innovation:** On this shared and democratized digital space, sandboxes for safe innovation can be created. A sandbox is a controlled environment where potentially risky systems can be tested without causing harm to the public. Sandboxes allow developers, startups

and researchers to test and refine open AI models safely. Because open models are transparent and easier to review, regulators can guide and approve them more effectively.

4. **Incentivize Open AI Models:** Beyond compute, governments can offer grants and incentives to encourage the development of open AI models. These could include prioritizing companies with open AI models in government contracts or providing risk support through multilateral development banks. Such policies help mobilize dormant capital and channel it into sovereign technology development.

5. **Launch Publicly Funded Open AI Missions:** Countries should establish dedicated missions that support the training, finetuning and deployment of open-source AI models, especially for public good applications such as climate change. This can be integrated into a National Digital Public Infrastructure Mission. Such a mission would promote both open data and open AI models as common rails or highways on which innovative solutions can be developed and scaled. This also helps countries build indigenous technology platforms on top of DPI.

6. **Promote Public Participation:** The advantage of open AI models is that anyone can contribute to their development and improvement. By turning open AI models into a citizen-driven movement, AI can reflect a far greater diversity of human experience. Proprietary models often limit this level of inclusion. Therefore, mass AI-literacy and inclusion drives should be undertaken so as to ensure that all citizens become part of their country's incredible AI journey.

7. **Boost Confidence through Auditability and Accountability:** Governments should implement transparent auditing frameworks that monitor bias, privacy and ethical impacts in open AI models. Public reporting and independent evaluations can strengthen trust by making it clear how these systems perform in real-world conditions. Because open models are inspectable, regulators, researchers and civil society can examine them more thoroughly, identify risks earlier and ensure that necessary safeguards are in place. This creates a shared sense of accountability and builds confidence in the safe deployment of AI for public use.

8. **Focus on Accelerating Cross-border Data Flows:** Cross-border data flows are essential for Climate AI because climate risks do not stop at national borders. Secure international data sharing enables AI models to combine satellite, meteorological and ground-level data from multiple countries, improving accuracy of climate solutions. The next step is federated and interoperable data frameworks that preserve data sovereignty while allowing real-time regional insights to be shared. This enables coordinated adaptation and mitigation responses rather than fragmented national action.

9. **International Organizations as Enablers:** International organizations should help countries and companies assess and manage risks around open AI while encouraging governments to adopt it for public benefit. They can also support compute and cloud infrastructure, offer financial incentives and promote harmonized standards for cross-border data flows so that privacy is protected while collaboration is encouraged.

10. **Proactive Role of Big Tech Companies:** Big technology companies can recognize the strategic and societal benefits of open AI and contribute by embracing transparency, sharing expertise and collaborating openly. They can gain access to national datasets and, in turn, use those insights to build freemium models that provide essential services for free and advanced capabilities for a fee. This creates a mutually beneficial dynamic that supports both public interest and business growth.

11. **Establish Monitoring and Impact Assessment:** Governments should continuously monitor and track the innovation, deployment and societal impact of the open AI ecosystem. These insights can guide regulatory adjustments, risk mitigation strategies and long-term planning in real time.

12. **Treat Data, Compute and Talent as Critical Security Assets:** Countries must recognize that datasets, compute capacity and skilled talent form the foundation of AI power. These resources should be protected, expanded and treated as critical infrastructure for national competitiveness and security.

CRAFTING A SYSTEMATIC WAY FORWARD

If open data forms a key pillar of DPI, then open AI models serve as a fundamental component by providing accessible, community-driven AI infrastructure that can be adapted and extended across sectors. These two pillars complement each other: open data delivers the raw material, and open AI models provide the tools built on that foundation. An open-source AI ecosystem acts as a powerful equalizer by democratizing access to advanced AI technologies and preventing monopolization by a few corporations or countries. A 2024 Harvard Business School study demonstrates that open-source technology significantly reduces software development costs, showing that companies using open-source spend up to 3.5 times less than they would if developing software from scratch.[12]

Innovations such as green data centres, open data for Climate AI, and open AI models are not merely solutions to AI's challenges—they represent transformative opportunities that stimulate innovation across climate and other sectors. However, these account for only about 30 per cent of what is needed to fully unlock Climate AI-fuelled innovation that is context-aware, inclusive, equitable and catalytic. The remaining 70 per cent consists of essential levers—policy and finance—which are critical both to realizing these opportunities and to supporting the broader ecosystem. Combining these levers with foundational innovations will create the conditions necessary for sustainable and impactful Climate AI progress.

Given the multidimensional complexity and the deeply intertwined nexus between climate and AI, it is crucial to take a systematic and structured approach to deploying these levers and advancing solutions. Without visualizing the big picture and aiming for coherence, efforts risk becoming fragmented and ineffective.

The next chapters discuss what it will really take to ensure that AI becomes a powerful force multiplier in our collective fight against climate change.

16

A SYSTEMS LENS

*A Pragmatic Policy Framework for Approaching
the AI–Climate (or Any Other) Nexus*

If you have reached this point in our book, congratulations. We are truly grateful that you have journeyed with us through the layered conversations on climate change, AI and the ways in which they shape and are shaped by one another. These themes are complex, filled with theory, reality, interdependence and challenge, and we have tried our best to make them engaging and accessible. What we hope made this journey even more meaningful was seeing some of these ideas through the lives of Asha, Sophie and Ori. If their stories added warmth or a spark of reflection along the way, then we feel our purpose has been fulfilled.

FROM SILOS TO SYSTEMS

Solutions to many of the world's most persistent challenges do not fail for lack of ideas, intent, or resources. They fail because they are pursued in isolation, while the ecosystems within which they operate remain fragmented. Infrastructure advances without institutional reform. Technology outpaces regulation, while knee-jerk regulation, in turn, stifles innovation. Incentives reward short-term gains even as long-term risks quietly accumulate.

Across the Climate AI solutions explored in earlier chapters, from forecasting floods and tracking wildfires to predicting crop health and monitoring global emissions, one pattern is consistent: impact is maximized when these solutions operate as part of a connected, end-to-end ecosystem. This is why we have repeatedly emphasized the importance of data integration, not as a technical detail, but as a foundational enabler of scale and effectiveness.

Moreover, AI's own dilemmas—its growing carbon footprint, the lack of accessible, high quality data, and the dominance of closed AI models—have all been framed in this book not merely as constraints, but as opportunities to design a blueprint for the future.

Yet for Climate AI solutions to deliver real-world impact, and for the challenges identified above—AI's carbon footprint, limited access to high-quality climate data, and the dominance of closed AI models—to be addressed meaningfully through solutions such as green data centres, democratized climate data, and open, accessible AI models, neither the solutions nor the systems that support them can operate in isolation. Their effectiveness depends not only on technological capability, but on whether they are embedded within ecosystems that align stakeholders, data, institutions, infrastructure, finance, and governance.

The task ahead is not simply to generate more solutions in isolation, but to design the systems required to integrate them and sustain impact over time. And if AI's dilemmas are indeed opportunities, the defining question becomes whether we are prepared to build the ecosystems needed to realize them.

So far, we have presented many ideas which could be central towards the creation of a blueprint that can transform Climate AI into something practical, impactful and inclusive.

But how do we actually construct this blueprint? How do we ensure that it achieves the impact that it is supposed to? How do we turn this vision into reality, and what should be the north star that guides us, helping us cut through complexity? This requires a framework, and that requires a vision.

Before we turn to the next section, it is worth pausing for a brief disclaimer, one that appears more than once in this book for good reason.

Our aim is to speak to a wide and diverse audience. Some readers may be familiar with systems design and find parts of this discussion deliberately high-level, while others may be encountering ideas that follow for the first time and appreciate a gentler introduction. That balance is intentional. At times, we zoom out for those who are accustomed to working deep within systems, and at other times we zoom in for those who may be viewing them from a distance. But regardless of where one starts, the central message of this chapter remains the same: effective climate action, the responsible channeling of AI toward it, and indeed the promise of Climate AI solutions, all depend (spoiler alert) on an ability to understand how the entire ecosystem works, how its parts interact, and where change can be most meaningfully applied.

A SYSTEMS LENS

Cutting through complexities often requires a systematic approach. Everything around us—in fact, we ourselves—represent a system. The human body is a remarkable system made up of different parts, each with its own role. These components work together to keep us healthy and functioning. Governments, private companies, multilateral institutions and startups all operate as systems. Each of them is a standalone system, and the way in which they interact with each other also represents a system.

Let us take a startup, for instance. Think of it as a system that includes founders, employees, technology, funding sources, customers and partners. Each part plays a distinct role, and all must work together in harmony to turn an idea into a successful business. In fact, every employee brings a unique skill set: one person may excel at coding; another at raising capital; someone might be skilled at forging partnerships, while another has a deep understanding of the policy ecosystem within which the startup operates.

Let us imagine that this startup now receives an incentive through a government policy seeking to incentivize Climate AI. It then develops an AI solution that encourages consumers to install solar panels by helping them understand the true solar potential of their rooftops and the money that they could save, should they install solar panels. Once consumers install the panels, their electricity bills are adjusted based on the energy they generate versus the energy they consume, with the electricity provider billing them accordingly. The interaction between the startup, the government, the consumers and the electricity provider, forms yet another interconnected system. The startup, which is also a system in its own right, is also part of this larger system.

At its simplest, systems thinking is a way of understanding how different parts of a whole interact with one another over time. Rather than examining individual components in isolation, it focuses on relationships, patterns, and dynamics. It recognizes that outcomes are shaped not just by individual decisions, but by feedback loops, incentives, delays, and information flows. In complex, fast-evolving domains such as climate change and AI, where actions in one part of the system often ripple across many others, this perspective is crucial. Of course, beyond this simple explanation, system design is a comprehensive discipline in its own right and we do not intend to cover its complexity in this book. Over several decades, researchers and practitioners have explored how complex systems behave, why unintended consequences arise, and how long-term outcomes are shaped by structure rather than intent alone. This book engages with these ideas only at a high level. Some readers may already be deeply familiar with systems thinking, while others may have little interest in exploring it further. For those who are curious, however, there exists a rich and well-established body of work shaped by thinkers such as Jay Forrester, Donella Meadows, John Sterman, Russell Ackoff, Peter Checkland, Herbert Simon, and Stafford Beer, whose ideas have influenced how governments and organizations approach complexity across domains.

Looking through a systems lens allows us to better understand the strengths and weaknesses of each piece of the puzzle and to put them together in a way that ensures a complete picture. Most importantly, it allows us to proceed with the understanding that our problem or the solutions to it are not a singular piece; it requires multiple pieces to come together.

THE FRAMEWORK: 6C MODEL

Looking at challenges through a systems lens also means paying attention to a few foundational elements that shape how outcomes emerge. These include the actors involved such as governments, firms, citizens, financiers, and institutions; the rules and incentives that guide behaviour; the physical and digital infrastructure that enables or constrains action; and the information flows that determine who knows what, and when decisions are made. Over time, these elements interact through feedback loops, reinforcing certain behaviours while discouraging others. When these components are poorly aligned, even well-designed initiatives can struggle. When they reinforce one another, progress can accelerate in ways that are difficult to achieve through isolated action.

We now present a simple, all-encompassing framework to design solutions—whether policies, strategies or interventions—that can help seize opportunities brought about by Climate AI. This framework ensures that our previous ideas, along with many more like them, are identified and implemented using a pragmatic and inclusive approach to maximize impact.

1. **Chart:** This step involves clearly defining what you want to achieve or the problem you want to solve. It means understanding the whole ecosystem—the stakeholders involved, how they complement each other, and where gaps exist—and mapping it all together, like a detective connecting clues on a board. For example, if you are an IT minister advancing a Climate AI policy, you first assess AI and climate ecosystems independently. Your country might have a vibrant AI ecosystem with strong research and startups but limited climate action, meaning you will need to develop the cleantech side to unlock the full potential of Climate AI. Alternatively, both areas might be strong, requiring a focus on maximizing their synergy, or both could be nascent, requiring foundational work. You then identify stakeholders, starting with your own peers in the finance and environment ministries, to private companies, startups and multilateral banks, and visualize their roles, interactions and opportunities for stronger partnerships and coordination.

2. **Consult:** With so many stakeholders and interconnected factors, the solution must impact the whole ecosystem. To achieve this, it is essential to not only consider the views of all

stakeholders, but also to involve them throughout the entire process: from brainstorming and solution development to implementation.

3. **Converge:** In this context, convergence has four key steps. Firstly, it entails bringing together all the stakeholders you have consulted. Secondly, it involves identifying how their capabilities complement each other. For example, one startup might excel at developing and synthesizing a new material that enhances rock weathering to capture carbon, while another may have created an AI algorithm able to analyse thousands of compounds quickly and recommend even more efficient materials that can accelerate rock weathering. Convergence is about enabling those complementarities, bringing them together to maximize impact. Third, convergence means aligning their aims towards a common goal, in this case accelerating AI enabled solutions to combat climate change. The final step is to translate these connections and insights into clear action items, defining who is responsible for what, what needs to change, and how to ensure the overall process moves forward smoothly.

4. **Collaborate:** Once the stakeholders are converged, it is vital to ensure they collaborate not only on building solutions but also on implementing them. This cooperation helps turn ecosystem participation into tangible ecosystem impact.

5. **Cascade:** Cascading refers to scaling up and multiplying the impact of collaborative solutions. It involves spreading successful innovations beyond initial pilots to reach wider

users and sectors. This expansion often requires coordinated efforts and resources from multiple actors. Enablers such as public policy, finance, research and development, infrastructure development, and capacity building are essential to support this growth and sustain impact over time. Without these enablers in place, even the best plans can falter. Identifying and strengthening these foundational elements is crucial to transforming ambitious goals into tangible impact.

6. **Circulate:** This step entails bringing circularity across the entire process: charting, consulting, converging, collaborating and cascading. This needs to happen repeatedly in a cycle supported by continuous feedback. This ensures the system can learn and adapt over time. For example, in the case of AI and climate, as new technologies emerge or climate impacts change, stakeholders should come back together to update their goals, share new insights and adjust their strategies. This ongoing loop keeps efforts relevant and effective as the situation evolves.

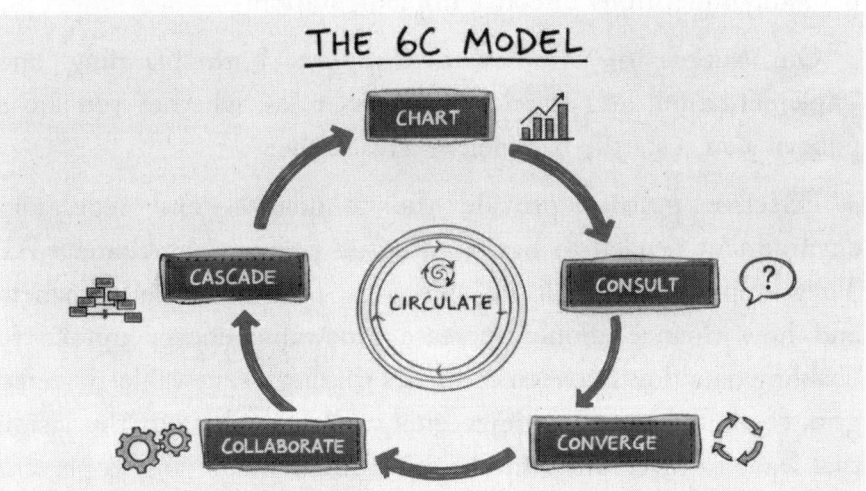

Interestingly, while we present this simple approach in the context of Climate AI, it can be applied universally across a wide range of challenges and opportunities.

The 6C model we have outlined may appear intuitive or even obvious at first glance. You might wonder why it needs to be stated at all. The reason is simple. In the rush to adopt new technologies, and in the heat of global competition, the fundamentals are often forgotten. As authors who have worked across government and the private sector, we have seen this repeatedly. Sometimes it is essential to pause, step back and return to first principles. The basics endure, and they work, especially when the challenges we are trying to solve are long-term in nature.

THE VISION: EFFECTIVE POLICY DESIGN

A robust planning process is one of the most critical steps in any solution, policy, strategy, or blueprint. Unfortunately, planning is often rushed or carried out in silos, especially when dealing with rapidly evolving technologies highlighted throughout this book. There is a strong desire to join the movement quickly, but inadequate planning undermines effective implementation.

Our simple 6C framework supports both planning and implementation, and is relevant across roles, whether you are a policymaker, a startup founder, or a researcher.

Effective policies provide the foundation and regulatory environment needed to harness the full potential of Climate AI. They help balance multiple objectives, from determining where and how finance should increase renewable energy uptake to enabling data flow between countries sharing a renewable-powered grid, so that AI can optimize energy. Policies lay out the vision that leads to diversification of critical mineral sourcing to prevent extreme weather events from bringing a nation to its knees. They lay out what amount of finance is needed, what incentives and subsidies should be offered, and who should be eligible. Essentially, a policy is a vision document with teeth.

When designed well, the 6C approach described here does more than improve coordination or policy outcomes. It reshapes incentives and creates strategic opportunities for the actors operating within those systems. The 6C framework shows how ecosystems can be deliberately designed, governed, and scaled. But systems do not operate on their own. Their effectiveness ultimately depends on how institutions and firms respond to the structures, signals, and incentives they create.

Viewed through a systems lens, the way firms engage with public priorities can be understood through a simple three-part logic we refer to as the 3C: Convergence, Collaboration, and Credibility. It retains Convergence and Collaboration from the 6C framework, albeit in a different context, and introduces Credibility as a new element.

As climate action and AI-driven transformation increasingly shape public policy, firms face a choice. They can treat these policy priorities as external constraints to be navigated through compliance, or they can allow their strategies to converge with them in a more deliberate and forward-looking way. When corporate objectives converge with public goals, space opens up for deeper collaboration, enabling firms, governments, and other institutions to combine complementary strengths in addressing challenges that are too complex for any single actor to solve alone.

Over time, such collaboration builds credibility. Firms that consistently align actions with public priorities and national and international goals can deliver tangible outcomes, potentially earning trust with regulators, partners, and communities. Just as importantly, collaboration around shared public priorities often spurs internal innovation, as teams are pushed to work across silos, recombine capabilities, and develop solutions that can evolve into scalable products, services, or platforms. In many cases, what begins as alignment with a public objective can give rise to new offerings that are deployable well beyond the original context. In this sense, the 3C operates as a reinforcing loop within the broader system: convergence enables collaboration, collaboration builds credibility, and credibility makes further convergence and cooperation both possible and sustainable.

If the 6C lens helps us understand how systems can be deliberately designed, and the 3C sheds light on how firms engage within those systems, the remaining task is to ensure that policies and strategies are shaped in ways that make this interaction work in practice.

But what should be the hallmark of a good blueprint, policy, strategy, or solution? What should we aspire for these policies to embody, or represent? Here are a few ideas to think about.

1. **Practical vs Prescriptive:** Any policy or strategy, especially one that seeks to address or capitalize upon AI, climate and their nexus should be realistic and actionable, as opposed to being prescriptive. This would be possible when there is an understanding of the status quo, the challenges, the opportunities, and both the planning and implementation is done taking all stakeholders on board collaboratively. It should identify synergies across different components of the ecosystem, map key stakeholders, diagnose barriers and opportunities, and clearly set out priority action items along with defined timelines.

2. **Horizontal vs Vertical:** While individual sectors and ambitions do need strategies and policies that go deep into

the domain, when it comes to the complexities with Climate AI, given the multidimensionality, interdependencies and complexities that we have discussed, it would be helpful to take a horizontal approach. This simply means that AI and climate both cut across all sectors, and therefore any strategy that explores this nexus should take an ecosystem approach rather than a siloed one. It would also help to link any new policy and strategy to those that exist, identifying how they can act as force multipliers, or what needs to change in them to accelerate action.

3. **Dynamic vs Static:** Both AI and climate are advancing at an unprecedented pace, presenting significant opportunities alongside complex challenges. As a result, any strategy or approach must be inherently dynamic, continuously adapting to evolving realities. This requires mechanisms that ensure the availability of real-time information and provide the flexibility to continually refine and evolve decisions, policies and strategies.

4. **Proactive vs Reactive:** Such an unprecedented pace also means that we should not keep waiting for the effects of climate change to manifest and then respond with knee-jerk reactions. The problem with this approach is that solutions developed in haste tend to be rushed and often fail to take an ecosystem approach into account. This means that the solutions are likely to be less effective and short-term, which is certainly not desirable in this context.

5. **Risk-tolerant vs Risk-averse:** The pace of innovation must keep up with, or ideally exceed, the accelerating progression

of climate change. Effectively addressing this challenge requires radical innovation and a courageous mindset that empowers innovators to take substantial risks, including the possibility of failure. To nurture such groundbreaking efforts, supportive mechanisms such as guarantees, financial loss protections, and risk-sharing arrangements are essential to alleviate concerns and encourage bold action. Traditional approaches used in conventional sectors are insufficient, and a fundamentally new framework is necessary to unlock transformative breakthroughs in climate solutions.

6. **Contextual vs Imported:** As is the case with sovereign AI, it does not make sense to import something as-is, from a completely different context. It may never prove to be effective. At worst, it could be counterproductive. There is no one-size-fits-all framework for Climate AI, and therefore, policies and strategies must be tailored to the contexts in which they are operating. Would it make sense to have a policy that promotes large-scale electric vehicle adoption in a country which has an underdeveloped electricity grid and limited charging infrastructure? Tailoring approaches to local environmental, social and economic conditions is essential for effectiveness and sustainability.

7. **Co-Created vs One-Dimensional:** One of the most important points emphasized in this chapter is the need for an inclusive and collaborative approach that engages all stakeholders from the outset. It is imperative to involve all stakeholders in both policy formulation and implementation. Their engagement at the planning stage will greatly enhance effective execution. The focus should not only be

on consulting stakeholders but also on co-creating policies and strategies with them, giving them a stake in the system to foster a sense of empowerment. This is especially critical because Climate AI requires cross-sectoral collaboration.

8. **Globally Integrated vs Isolationist**: This takes the need to co-create policies and strategies a step further by emphasizing that all policies aimed at climate action, artificial intelligence, and their intersection must be interconnected and mutually reinforcing. As we have discussed, climate is a global commons. It affects everyone and its impacts do not stop at national borders nor do its causes. Whether it is the necessity for data to flow seamlessly across borders or artificial intelligence-enabled optimization of power supply on shared grids between countries, all such challenges require an approach where nations do not ring fence resources or restrict collaboration. Instead, they must foster open cooperation, shared governance frameworks, and interoperable systems to collectively address these transnational issues effectively and equitably.

9. **Evidence-Based vs Assumption-Driven:** In an ideal world, policies and strategies aimed at Climate AI should be based on data and grounded in evidence rather than assumptions. While it is true that the pace at which climate is progressing means we may not always have the evidence needed to make every single decision, and we certainly are not advocating for evidence to become red tape, evidence-based approaches do increase the likelihood of effectiveness. If anything, this should motivate us to begin collecting and making data publicly available on a massive scale, something we have been advocating for throughout this book.

10. **Motivating vs Mandating:** When it comes to policies and strategies aimed at Climate AI, while mandates are important, sometimes it makes sense to take an approach where stakeholders are actively encouraged and incentivized. Sometimes mandates become important, for instance, say data protection laws in the case of AI and net-zero goals in the case of climate, but at an ecosystem level, a balance of both approaches is necessary: mandates provide the regulatory foundation and guardrails, while incentives and encouragements drive broader participation, experimentation and scalable impact across diverse stakeholders.

CALL TO ACTION

If the 6C framework explains how systems can be deliberately designed to support Climate AI, the 3C framework shows how companies operating within those systems can treat regulation and public priorities not as constraints, but as opportunities to innovate, build trust, and create long-term value. The vision for policy and strategy outlined above is intended to make this interaction work in practice, shaping incentives, reducing friction, and enabling coordination at scale.

The underlying mantra to maximize effectiveness is now simple:

1. **Build Systems Capability, Not Just Technical Expertise:** As climate action and AI deployment become more complex, there is a growing need to strengthen systems capability across governments, companies, and institutions. Technical expertise in AI, energy, or climate science is necessary, but no longer sufficient on its own. Decision-makers, regulators, entrepreneurs, and practitioners must be equipped to

understand how incentives, infrastructure, data, finance, and governance interact within broader ecosystems. This calls for targeted training in systems thinking and systems design, and their inclusion into public administration, corporate leadership programmes, and technical education. Building this capability, using the 6C model as reference, can help actors move beyond siloed decision-making, identify priorities more clearly, and design interventions that are resilient, adaptive, and scalable over time.

2. **Identify and Act on Leverage Points Within Climate–AI Ecosystems:** One particularly useful idea from systems thinking is that of leverage points. Leverage points are places within a system where relatively small, well-designed interventions can lead to outsized and lasting change. They are rarely found at the level of visible outcomes, such as emissions figures or deployment targets. Instead, they tend to sit deeper within system structures, especially in how information flows, incentives are set, and decisions are coordinated.

   In the context of AI and climate, this book has already explored two powerful examples of such leverage points: democratized climate data and open, accessible AI models. As discussed in earlier chapters, improving access to high-quality, interoperable climate data can unlock innovation across mitigation, adaptation, and resilience by lowering barriers for researchers, startups, and policymakers alike. Similarly, open and sovereign AI models can reduce dependence on a handful of actors, enable context-specific solutions, and accelerate learning across regions. These interventions do not directly reduce emissions on their own. Instead, they reshape the system by changing who can participate, how quickly solutions can spread, and how effectively knowledge circulates. When leverage points such as these are identified and activated, they can trigger cascading benefits, allowing multiple solutions to scale simultaneously and making the overall system more adaptive, inclusive, and effective.

3. **Design Policies and Strategies to Enable Convergence, Collaboration, and Credibility:** Policies and strategies aimed at the AI–climate nexus should therefore be designed to enable convergence between public goals and private strategy, facilitate collaboration between companies and governments, and help build long-term credibility to foster mutual trust. This means moving beyond narrow compliance frameworks toward policies that reduce uncertainty, reward long-term investment, and create space for firms to co-create solutions alongside governments. Similarly, for firms, this means viewing policies, priorities and regulations not merely as compliance obligations, but, where relevant, as opportunities for innovation.

   When policy design reinforces these dynamics, alignment becomes durable, collaboration becomes scalable, and innovation can translate into both societal impact and commercially viable offerings.

Ultimately, however, even the best-designed systems and policies fail without effective enablers, and the most decisive of these is finance.

17

FUNDING THE FUTURE

Financing AI-Driven Climate Innovation and Action

One of the most important levers to support any solution, scheme or project is finance. Today, finance, or rather the lack of it, is one of the biggest challenges when it comes to scaling up climate action at the global level. At the same time, investors across the world are bullish when it comes to financing AI, driven by its promising potential and also its ability to generate significant economic value.

Transformative innovations, especially those at the intersection of AI and climate action, simply do not move from concept to real-world impact without sustained investment at every stage. In this chapter, we will explore the importance of financing throughout the Climate AI lifecycle, identify key barriers, discuss available funding sources, and finally put forth a few ideas.

FINANCING ACROSS THE CLIMATE AI LIFECYCLE

Setting aside large companies that already have access to capital and a willingness to take risks, it is important to recognize that some of the most groundbreaking Climate AI solutions will come from startups, researchers and innovators. As discussed in Chapters 14 and 15, Climate AI development can be accelerated by democratizing access to open data. This accessibility will empower startups, researchers and academic institutions to begin innovating and creating impactful technologies. However, democratization alone is not enough. Finance is crucial at every stage of the Climate AI development—let us see how.

Outside his work at the DFTCCM for Mayor Ashman in Climaville, Ori nursed an idea of his own. He wanted to build a small startup that could help people in low-income countries earn money by taking climate-positive actions in their daily lives.

His concept was simple but powerful: an app that would allow farmers with small landholdings to get paid by global companies for the carbon they sequester through practices like agroforestry, biochar, soil regeneration and better water management. Companies would get verifiable offsets, and farmers would receive a new income stream for climate-friendly practices they were often already doing, but rarely recognized for.

Like all climate technology, the journey began with research and early innovation. Ori spent late nights applying for grants and small innovation awards, hoping to gather enough environmental and agricultural data to train early AI models. He needed to demonstrate that his idea was technically viable: that a smartphone camera, a few soil measurements and satellite imagery could work together to estimate carbon capture accurately enough for companies to trust.

The next hurdle was testing and validation. Even though his models looked promising on paper, he had to test them in the real world. That meant travelling to farms, validating soil readings and understanding how the app worked in places with erratic connectivity. He quickly learned that farmers had different practices, different tools and different levels of digital confidence. Several pilot tests stalled because he simply did not have the funding required to run proper field trials, train local partners and build trust with communities.

Expansion and integration proved even more challenging. To scale the platform, Ori needed to build robust systems that could integrate with carbon registries, agricultural extension networks and even financial inclusion programmes so that farmers could receive payments instantly. He also needed to train people who could help farmers adopt the app, update local soil baselines and maintain the digital infrastructure. Deploying something like this at scale demanded far more capital than he had imagined. It was not just writing code; he needed cash too.

And even if he could pull all that together—and he remains confident that he would—sustained operation and evolution required continuous funding. His app would need constant improvement, new AI capabilities and the flexibility to adapt to local ecosystems. He imagined a world where the platform started with grant support but eventually became a self-sustaining business: farmers subscribing for advanced advisory services, insurers paying for better risk modelling, companies purchasing high-quality domestic offsets. But all of that required time, talent and patient capital.

THE BARRIERS

While finance is essential, innovators face several significant challenges that act as barriers to progress. Addressing these obstacles is key to unlocking the transformative potential of Climate AI. Let us explore some of these challenges, presented in a simplified way for our readers who come from diverse backgrounds, countries, cultures and who bring varied experiences, expertise, passions and curiosities.

1. **Profitability Gaps:** Market-driven finance tends to prioritize quick and profitable returns, but many essential Climate AI solutions, such as disaster alert systems or open environmental data platforms, provide significant social and environmental benefits without immediate financial gains. Because these solutions do not generate fast profits, they often struggle to attract sufficient funding. This mismatch creates a barrier that leaves important projects focused on adaptation and resilience underfunded and underdeveloped.

2. **Capital-raising Challenges:** Startups developing these solutions face extended timelines to generate returns, complex regulatory environments, and uncertain market demand that is often driven primarily by government agencies (which is a less profitable scenario). This combination of factors makes them appear risky to conventional investors and lenders.

3. **Fragmented Landscape and Rigid Criteria**: The climate finance landscape is often complicated by stringent and varied requirements, fragmented across different sources and regions. This creates significant hurdles for innovators who must navigate diverse eligibility conditions and deal with rigid standards that simply do not take into account the characteristics of frontier technologies. By virtue of the complexities these solutions seek to address, drawing on an interplay of AI and climate, they sometimes do not fall neatly into existing funding categories, making it harder to secure support.

4. **Global Inequities:** Developing countries, overall, face persistent challenges in accessing adequate climate finance due to limited domestic capital, higher perceived risks, and pressing priorities for basic infrastructure, health and education, which take up a large chunk of funds. Despite

pledges of substantial international climate funding, the actual amounts delivered often fall short of promises. We have touched upon this extensively in Chapter 5, which is focused on rethinking climate equity and how this is a shared yet differentiated burden for countries. The finance from the developed to the developing world remains abysmal. Moreover, large portions of this funding come as loans instead of grants, increasing debt burdens for developing countries and reducing the available resources for local innovation and climate adaptation efforts that are urgently needed.

5. **Perception and Optics:** Climate investment is sometimes perceived as a non-essential or philanthropic expense rather than a core economic opportunity. Many investors shy away from technologies that are unproven, face scientific uncertainties, or rely on evolving policy frameworks. This perception is even more pronounced during times of economic uncertainty or shifting global climate priorities, leading to missed chances for funding transformative innovations that can accelerate climate action.

CORE FINANCE STREAMS

Overcoming these barriers will require concerted, concentrated and continuous efforts. We cannot claim that we can solve all of them, but we do have some simple and straightforward ideas around how the future of financing for Climate AI could look like. But before we explore those strategies, it may be useful to have a quick look at the current sources of funding. For clarity, we have categorized these fund sources into five main streams.

1. **Public Sector Commitments:** This includes financial resources that come directly from national and local governments. These may take the form of public investments, subsidies for clean technology, regulatory incentives and national budget allocations for climate innovation. Governments also play a central role in mobilizing and distributing international climate finance into domestic projects. Examples include India's National Mission on Green Hydrogen, and the European Union's Innovation Fund, which supports large-scale climate technologies. Many countries also run public climate innovation funds such as the UK's Net Zero Innovation Portfolio.

2. **International Climate Finance:** These are funds provided bilaterally through multilateral climate funds, development banks and global climate mechanisms. Much of this is guided by the principle of 'common but differentiated responsibilities' where developed countries support developing countries. Key examples include the Green Climate Fund, the Global Environment Facility, the Adaptation Fund, the Climate Investment Fund as well as funding from multilateral development banks such as

the World Bank or the Asian Development Bank. Bilateral climate finance from Germany's International Climate Initiative, and Japan International Cooperation Agency (JICA) are also some of the many such contributors.

3. **Self-sustaining and Market-based Investment:** This includes venture capital, private equity, debt financing, blended finance, startup investment and revenue-driven business models. Companies and entrepreneurs raise capital from the market and also design solutions that eventually pay for themselves through mechanisms such as subscription fees, pay-for-performance arrangements, energy-as-a-service models or the sale of carbon credits. Examples include venture funding from firms like Lowercarbon Capital, Breakthrough Energy Ventures and Energy Impact Partners. Market-based climate instruments such as the Voluntary Carbon Market, legacy compliance mechanisms such as offsets generated under the Clean Development Mechanism and agriculture-based carbon programmes like Indigo Carbon also fall in this category.

4. **Philanthropic and Development Assistance:** These funds come from international aid agencies, charitable organizations, philanthropic foundations and impact investors. They are essential for filling early-stage gaps that neither governments nor markets address quickly enough, especially for community-based solutions and high-risk climate innovation. Examples include the Bezos Earth Fund, the Rockefeller Foundation, the Gates Foundation and the Hewlett Foundation. Philanthropy has been particularly important in climate resilience, early

warning systems and community-level adaptation where market incentives are weak.

5. **Innovative and Emerging Financing Instruments:** An increasing portion of climate finance now comes from innovative tools that blend public, private and philanthropic capital. These include blended finance platforms (which use concessional or philanthropic funds to pull in larger commercial investment), sovereign green bonds (government bonds issued to raise money exclusively for environmental or climate projects), resilience bonds (insurance-linked instruments whose pricing or payouts depend on achieving climate resilience outcomes), sustainability-linked loans (financing where the cost of borrowing changes based on meeting sustainability targets), and debt-for-climate swaps (agreements that redirect debt repayments towards climate action). These mechanisms help attract private investors by reducing risk and improving the predictability of returns for climate solutions. Governments are also turning to the monetization of brownfield assets, using the value of existing operating infrastructure to fund the next generation of green and climate-focused projects.

BIG-PICTURE IDEAS

Now that we know the barriers and the broad streams of finance that could be tapped into, it is perhaps time to shift our attention from analysis to action by exploring some ideas that could potentially help fund Climate AI. We call these 'ideas' rather than 'solutions' because there are already thousands of detailed books, reports and resources available on climate finance and innovation, each giving out prescriptive solutions. Our goal is to just zoom out, and focus on simple but practical concepts that can drive significant impact, when planned and executed thoughtfully and robustly (making Chapter 16 on systems lens more relevant than ever). We also do not aim to majorly replicate existing recommendations on climate finance. You can easily find extensive resources and expert advice through platforms like Perplexity, ChatGPT or DeepSeek. Instead, we offer broad ideas designed to inspire strategic thinking and action at scale.

1. **Piggyback on AI for Climate AI:** Even though global uncertainties shroud climate finance, and delivery on climate finance remains poor, the world remains bullish on AI. Investment in AI is growing across industries. Such investment helps build the overall AI ecosystem, including technology, talent and data infrastructure. Climate AI can benefit by leveraging this growing AI foundation. Strategically aligning Climate AI efforts with the wider AI investment trend offers a promising way to accelerate solutions for climate challenges. Hence, innovators should piggyback on AI to help accelerate Climate AI, including by positioning their Climate AI solutions as AI solutions first.

2. **Shift the Narrative:** From an optics perspective, we must shift the narrative surrounding climate and its economic impact. Discussions about AI often highlight its potential to boost GDP by significant percentages. In contrast, conversations about climate change typically focus on the damage it causes and the negative effects on GDP. This framing undermines confidence and support for climate action. We need to reframe the dialogue to emphasize the positive economic contributions of climate solutions, such as the percentage by

which climate action can increase GDP, the new jobs created, cost savings from energy efficiency, and new markets for green products. Although this may seem like a subtle change, it is crucial for building optimism and attracting investments, specially from the private sector.

3. **Create Fund of Funds:** Countries should create a 'fund of funds' which consists of budgetary allocations from the government, grants from multilateral development banks, money from philanthropies and corporates willing to pump in their corporate social responsibility (CSR) funds. This fund of funds should be reserved for startups that are innovating on Climate AI (ideally, it should be extended to all things climate). The application process should be simple and there should be no rigid eligibility criteria. We must trust our startups. If their ambition is commendable, they should be allowed access to funds.

4. **Government Guarantee to Grow Innovation:** Countries should recognize and address the significant barriers before innovators when they seek capital from banks. While large corporations often secure substantial loans even when they default without severe consequences, small innovators struggle due to lack of proven technology and collateral. The government must act as their collateral, sending a powerful signal of support that can unlock vital funding and fuel groundbreaking innovation.

5. **Encourage and Incentivize Startup–Private Sector Links:** Private companies should be encouraged and rewarded for

working with startups to create Climate AI solutions, instead of relying only on their own teams. This approach brings fresh ideas and top talent to the companies while giving startups the funding they need to grow. By working together, both sides can help speed up the development of effective tools to tackle climate change.

6. **Reward Climate Co-benefits:** Countries should focus not only on incentivizing Climate AI innovators, but also on rewarding the broader economic benefits their work generates. They should create Climate Co-benefit Linked Incentive Schemes (CLI) that provide rewards to Climate AI solutions demonstrating clear economic spillover effects beyond their primary environmental goals. For example, an AI system designed to optimize clean energy use that also creates new local jobs or reduces energy costs for small businesses would qualify for such incentives. Similarly, a Climate AI tool that improves agricultural resilience while boosting farmers' incomes could receive additional support under this framework. This approach encourages innovation that drives both climate impact and tangible economic growth.

7. **Build Global Investment Portals:** Imagine an Amazon or eBay where investors browse and fund Climate AI solutions instead of buying products. Each listing describes the innovation's features, potential impact and progress, similar to product descriptions on e-commerce sites. This portal serves anyone interested in funding projects, offering a centralized showcase of innovation and making investment decisions easier through transparency and organization. Just as how countries create investment pipelines for infrastructure

projects to attract investors by presenting clear opportunities, this platform connects investors directly with scalable, impact-driven solutions. This portal should become the go-to destination for climate finance, and it should be promoted widely at events from COPs to Climate Weeks. This gives innovators global visibility, access to broader capital pools, and fosters collaboration across borders.

8. **Channelize Development Flows Globally:** Let us consider expanding the functionality of these portals. A major challenge with climate finance is that funds often flow to where visibility is highest rather than where the impact would be greatest. For example, does it make sense to finance additional renewable capacity in a country with a relatively clean grid already supplied mostly by renewables, rather than in one with a heavily coal-dependent grid? The key is to institutionalize this level of impact visibility on a global scale. Such a platform could be linked to the investment portal proposed earlier. Powered by AI, users could ask questions such as where to invest in solar energy in Africa to achieve the greatest impact. The AI would recommend the optimal country, allowing users to explore relevant innovators, projects and funding opportunities there, with the potential option to even zoom into city and village level, and do deep dives into financials.

9. **Monetize Data, the Freemium Way:** For innovators developing solutions for the public good, exploring alternative revenue streams is essential to sustain funding. For example, creating an AI-powered flood forecasting or wildfire tracking system provides early warnings that save lives and serve society.

However, the valuable data generated can also be monetized by selling it to businesses such as insurance companies, which can use the information to assess risk and improve their services. This could follow a freemium model where the public receives free alerts while commercial users like insurance firms pay for access to advanced data or enterprise features. This approach is similar to how map services operate, your Apple and Google Maps are free for you to use, but if food delivery and ride hailing apps use them, they have to pay a premium subscription.

10. **Deploy Blended Finance:** Blended finance can unlock Climate AI innovation by strategically combining public, private and philanthropic capital to reduce risk and accelerate deployment. A first-loss guarantee becomes especially consequential in the case of Climate AI, where projects often rely on untested data models, uncertain revenue streams and long validation cycles. By absorbing initial losses if a project underperforms, such guarantees make it possible for private investors to back Climate AI ventures that would otherwise appear too risky. For example, a startup developing ultra-local climate risk prediction tools could attract investment far more easily if the riskiest tranche were protected through a first-loss guarantee mechanism. Beyond guarantees, blended finance can mobilize concessional debt, catalytic equity, results-based grants and climate-risk insurance to create investment stacks that match the unique risk profile of Climate AI solutions. Another innovative approach is the creation of impact-linked finance for Climate AI, where the cost of capital decreases as the model's climate benefits improve, aligning financial performance with measurable adaptation

or mitigation outcomes. When layered thoughtfully, these instruments not only fill financing gaps but also crowd in private capital, de-risk experimentation and help transform Climate AI pilots into scalable national platforms.

11. **Evolve Megascale, Democratized Carbon Markets:** Launching carbon credit markets at a mega scale and making them accessible to individual users could transform climate finance. Today, carbon markets are largely limited to corporations and governments, with high entry barriers that restrict broader participation. By expanding access and allowing everyday users, startups and small businesses to participate, carbon markets could become far more vibrant and liquid. For example, a cement manufacturer in Croatia could offset its emissions by purchasing credits from a community-run clean energy or reforestation project in Nigeria. Such cross-border trading would increase liquidity and channel finance directly to places where emission reductions actually occur. A platform of this kind could also fund startups that deliver verifiable carbon reductions, supported by technologies such as blockchain and AI to improve transparency and trust. Success would require strong verification systems, simple user interfaces, regulatory alignment across countries and integration with financial services so that carbon credits can be traded as easily as any other asset. A more democratized carbon market would broaden participation and embed climate action into everyday economic activity worldwide.

12. **Monetize Underused Brownfield Infrastructure for Climate Action:** Governments hold vast amounts of underutilized land, warehouses, industrial sites and public

infrastructure that often sit idle without generating value. These assets can be monetized through long-term leases, public–private operating models or redevelopment partnerships, unlocking significant capital that can be directed into climate change management. This creates a win-win outcome. Infrastructure that was lying dormant is brought back into productive use, potentially stimulating local economic activity, while the upfront payments from private partners provide governments with additional resources to fund climate mitigation and adaptation efforts. By turning overlooked assets into revenue generators, countries can create a sustainable, domestic source of climate finance without increasing public debt.

13. **Use Finance to Reward Alignment, Not Just Performance:** Traditional climate finance often rewards narrow project-level outcomes. Instead, financing frameworks should be designed to reward alignment with broader public priorities, system integration, and long-term credibility. Capital can be structured to favour projects that co-locate with renewable energy, strengthen grid capacity, share data openly, or generate clear climate co-benefits for local communities. Over time, such incentives encourage firms to align strategy with public goals, collaborate across sectors, and invest in solutions that are durable rather than extractive.

14. **Finance System-Level Infrastructure, Not Just Standalone Projects:** Climate AI financing should increasingly prioritize system-level infrastructure rather than isolated solutions. This includes investments in green data centres, grid modernization, energy storage, data platforms, and shared

digital and physical infrastructure that enable multiple Climate AI applications to scale simultaneously. By funding these enabling layers, governments, development banks, and investors can unlock cascading impact, where a single investment supports many downstream innovations. This approach reduces duplication, improves capital efficiency, and strengthens the resilience of the overall ecosystem.

THE (SYSTEMIC) WAY FORWARD

Having broadly examined the importance of finance as a crucial enabler, the barriers that stand in the way, the existing sources of funding available, and a few interesting ideas to bridge the financing gap, we must also recognize the unique nature of Climate AI. This field depends on the interdependence of many actors such as governments, private companies, startups, financiers, research institutes, NGOs, end users, and sometimes entire communities. What also makes Climate AI different from other fields with similar interdependence is the urgency and complexity of the issues it tackles. These solutions must combine scientific data, policy frameworks, economic incentives, and social equity all at once. The long-term and unpredictable nature of climate impacts increases risks and financing challenges. It also requires bridging divides between public and private sectors, involving actors who usually work separately, including governments, communities, corporations and philanthropy. Because of this, Climate AI requires a multidisciplinary and multidimensional approach towards finance and implementation. For example, a private company is more likely to invest in a Climate AI solution benefiting an entire country if an entity such as a multilateral bank agrees to take on some potential losses to reduce the risk. This assurance encourages companies to commit resources. This risk sharing can only be enabled through the government of that country.

This is exactly why the ideas that we have suggested need to be seen through the systems lens we recommend in Chapter 16. In order to make them happen, we must plan exhaustively but also inclusively. We must bring together stakeholders, understand how they interact, map out synergies, leverage those synergies to drive action, and deliver together.

Finally, while it has not been included here as a recommendation since it is rather obvious, it is absolutely critical for the developed world to urgently step up and deliver on their climate finance commitments. International organizations must also sharpen their focus and drive concrete, coordinated action without delay. To hold all parties accountable, a global dynamic heatmap should be created to clearly highlight countries falling short of their responsibilities, shining a spotlight on where urgent progress is needed.

The time to move beyond promises and take decisive action is now.

18

CLOSING THE LOOP
Making Sense, Moving Forward, Together

In the complex, uncertain and multipolar world that we live in today, there are challenges and there are opportunities. As we have discussed throughout this book, the world is confronted with conflict, policy uncertainty, trade wars, a stagnation of economic growth and the devastating effects of climate change. Of these challenges, climate stands out because it not only causes damage to lives and livelihoods and collectively threatens the future of humankind, but it also does so by taking existing challenges and threats, and multiplying them.

For instance, climate change can cause shortages of resources, everything from food and water to minerals, not just hindering economic development but, in the worst case, fuelling conflict. Conflicts, as we are seeing today, threaten the global world order with everything from fuel prices to geopolitical equations. Climate can bring serious harm to the health and productivity of people, from exposing diseases of the past that lay buried beneath permafrost to vector-borne diseases arising out of floods.

It can also directly affect technology, which is very fundamental to the collective progress of humankind, by causing disruptions to supply chains of critical minerals. Recall Asha's country: a climate-driven drought-intensified local conflict, which in turn interrupted an international trade route carrying minerals essential for everything from electric-vehicle batteries to the data centres that power AI. These are not hypothetical risks; such disruptions are already occurring in parts of the world. Left unchecked, simultaneous breakdowns in multiple supply chains would pose a cascading, systemic threat to technology, the economy and society.

Climate change did not happen overnight. The developed world emitted carbon at a huge rate as coal and oil powered their industrial growth, which is one of the reasons why they are 'developed' today. In contrast, the developing world is still developing. Many of these countries have a delicate balance to strike between investing in lifting people out of poverty and focusing on immediate development priorities versus achieving net zero. This does not mean that they should not pursue net zero, and many of them are taking phenomenal leaps in this direction. It simply means that because they were not historically responsible for the current situation, they need assistance from the developed world, which reached its status through carbon-heavy pathways. Even today, emissions of developed countries remain exponentially higher as compared to the developing world. That is where climate finance and climate equity come into play.

Today, the impacts of climate change are being felt across the world and in our daily lives. Whether it is floods, wildfires, hotter summers, unpredictable monsoons, droughts, or frequent air turbulence, climate change is showing up across sectors and across countries. Still, these impacts are not distributed equally. Developing countries, which have contributed the least to climate change, are often the most vulnerable. Many people live on the margins, with fewer economic resources and safety nets, which makes them more affected when climate shocks hit. That is why carbon equity matters so much.

But there also lies an opportunity here. While it is true that developing countries have been the lowest contributors to emissions and their vulnerability is the highest, they also have the greatest potential, and that is to industrialize on low to no carbon pathways on the back of the cutting-edge technologies that are available today.

When the western world was industrializing, they did not have access to the kinds of technologies that are available today. Today, the world has wind, solar, geothermal and nuclear power. We are storing energy through gravity storage systems and large batteries. Carbon can be captured directly from the air and buried deep underground or used to manufacture everything from car tyres to materials for the aerospace sector. Green molecules such as green hydrogen, green ammonia and green methanol will fuel industrial decarbonization. New technologies are being scaled up through cutting edge innovations that will massively accelerate the transition to net zero. But again, this ties back to equity. The developing countries need finance and access to technologies, which are sometimes locked behind paywalls.

But the world today also has AI, which is growing at an unprecedented rate. It could come to represent humanity's greatest leap. Imagine complementing cutting-edge climate technologies with AI. It could truly become a force multiplier against climate change, which is a threat multiplier.

AI can help make power grids smarter, optimize transport and logistics, and generate insights that lead to greener products. It can cut methane emissions from agriculture, bring trust and transparency to carbon markets, and suggest the best mixes of materials for greener manufacturing. AI also has an incredible ability to analyse millions of material combinations quickly and simulate many 'what if' scenarios. This helps discover new materials that reduce carbon emissions and improve processes such as the production of green hydrogen and carbon capture. What once took years or decades of research can now be done in days or hours, speeding up the transition to a greener, cleaner future.

AI can help the world in adjusting to the new challenges climate impacts bring, like flooding, heatwaves, or droughts, so people and infrastructure can better withstand and recover from these shocks. For example, AI can analyse complex data about weather, land and infrastructure to predict risks, guide emergency planning, and recommend changes to make cities and farms more resilient. It can help design flood defenses, plan evacuation routes, and even forecast crop cycles to protect food supplies. By making sense of many different data points together, AI acts as a powerful tool for anticipating problems and supporting smarter decisions that save lives and livelihoods.

AI can also power smarter systems, improve governance, deepen behavioural insights, and sharpen financial decision-making for climate action. We have explored how AI can simulate policy outcomes, increase transparency, direct investments where they matter most, and encourage individuals to make lower-emission choices. It can help with everything, from real-time monitoring to long-term planning.

A simpler world would be easier to manage, but both AI and Climate have numerous complexities. While we are talking about AI helping aid climate action, AI itself contributes to emissions because of how much power and water it guzzles, with power consumption from data centres—the brains that power AI—

expected to rise and consume as much power as consumed by the entire country of India, the most populous country in the world.

Now, the demand for AI is clearly increasing, and so is the demand for data centres. Given the crucial role AI can play in climate action, this growth can be beneficial. However, it is essential that these data centres are green. They should be powered by clean energy, be highly energy-efficient, and not exacerbate water scarcity. Today, innovative data centres around the world are adopting efficient cooling techniques, co-locating with solar and wind farms, utilizing geothermal energy, and pioneering advanced methods such as using seawater for cooling or even placing data centres under the ocean.

Interestingly, many of these data centres are being built in developing countries, presenting another unique opportunity. Beyond building them as green and sustainable facilities, we have the chance to design them to contribute directly to the economic development of surrounding communities. The renewable energy powering these data centres can also supply local villages. Recycled wastewater can be used for cooling, providing communities with sewage treatment plants and treated water. Additionally, the heat generated by data centres can be harnessed to supply warmth to nearby residents. These models are indeed feasible, and they should become part of a mutually beneficial framework.

Win-win models like these require strong support. That is why their development must be incentivized, policy barriers removed, single window clearance systems established, and access to carbon-free energy made straightforward for them to procure. Such data centres would also benefit from improved trunk infrastructure in the areas where they are built, as well as from innovative financial mechanisms to assist with their construction. All of this can be enabled through dedicated data centres policy, international cooperation and public–private partnerships on their building, financing and operation. Finally, a key constraint is not the availability of renewable generation alone, but the capacity and intelligence of the systems that connect energy, data, and infrastructure. Designing data centres as integrated components of power grids, storage systems, and digital governance frameworks is therefore critical, both to manage scale and to ensure that their growth strengthens, rather than strains, underlying systems.

At the core of many Climate AI applications lies the central role of data, including both its availability and integration. This is what gives these solutions their power. Consider an AI model that predicts extreme heat. Based on such predictions, an AI system managing the electricity grid could optimize consumption to keep the grid stable as demand for cooling rises. Another AI solution could help the government identify which households are most vulnerable and most likely to be impacted, enabling timely cash transfers and preparation of medical infrastructure in those areas. A different AI tool could support builders of future data centres by testing different cement compositions to develop materials that keep facilities cooler than conventional cement. At the same time, an AI-driven platform could integrate data and insights from all these solutions, thereby assisting a city mayor in distributing budgets more effectively to ensure better heatwave management, both now and in the future. AI connects data across regions, sectors and systems to generate timely, tailored and actionable insights.

But for this to happen, data needs to be available. It needs to be collected and made accessible in an easy-to-use format, and not locked behind paywalls by the private sector. Since the impacts of climate extend beyond borders and require solutions built on international cooperation, such data must flow freely between countries. To this effect, we recommend digital public infrastructure or DPI for climate. DPI is gaining momentum globally, with India setting the gold standard. The country's payment system, which we have discussed comprehensively, demonstrates how the combination of digital identity, mobile numbers and bank accounts has become central to India Stack, a DPI that includes layers for payments, identity and consent. Instead of every company investing in and building its own data highway, they get onto a common one, which ensures solutions have wider applicability whilst also keeping data supporting Climate AI solutions democratized.

To make this happen, we recommend extending DPI to build climate and energy stacks. This will help with everything from facilitating cash transfers to people affected by climate disasters to enabling peer-to-peer energy trading, turning Climate AI into something with mass stakeholdership. Another core philosophy is to make all climate data open and accessible through national and international repositories. For example, building solutions for extreme heat requires access to temperature, grid demand, socio-economic and GIS data to help governments support the most vulnerable populations. Open data makes such comprehensive solutions possible. Just as climate is a global commons, so too must data become a commons. This calls for creating national Digital Public Infrastructure (DPI) missions, establishing data standards and sharing rules, winning private sector trust by demonstrating the win-win nature of these models, creating a global DPI body, mainstreaming DPI in national climate action plans, and leveraging AI to clean and organize vast amounts of data.

But this solves only part of the problem. How do we ensure that the Climate AI solutions being built are truly relevant to the specific contexts and countries where they operate? The answer lies in decolonizing AI through open and sovereign AI ecosystems. Open-source models like DeepSeek demonstrate that powerful AI can be developed transparently, affordably and collaboratively, enabling countries to retain technological sovereignty. This means building AI solutions tailored to local languages, cultures, governance, and climate vulnerabilities and avoiding the biases and limitations of closed models. To achieve this, we recommend treating open AI models as DPI—similar to open data—to democratize access and innovation. Governments and international agencies should host and provide computing resources for open AI, establish regulatory sandboxes for safe testing, incentivize open AI development, and launch publicly funded missions to support such efforts. Widespread public participation, transparent audits and strong international collaboration are critical to build trust and ensure inclusivity. Finally, big tech companies should embrace open AI principles and collaborate in creating a thriving open ecosystem. These steps create an AI environment where solutions are contextually appropriate, inclusive, and accelerate equitable climate action globally.

None of this will be possible without smart policies, robust regulations, access to climate finance, and meaningful partnerships. Solutions must not only be technically sound but also actionable and context-aware. As we have been able to fathom by now, when it comes to climate, AI and their nexus, there are too many interdependencies, too many interactions, too many complexities, but also too many possibilities. The ecosystem is multidimensional and multilayered, and therefore, the solution to advance Climate AI must be one that is multifaceted, multisectoral and multiscale.

This requires one to zoom out and really look at the big picture, and perhaps this is where we need to take a systems lens. Systems are all around us; our body itself is a system. A government, or a startup, a research institution, or a private sector company on its own is a system with many different moving parts. The way in which they interact with each other is also a system. Visualizing this helps, in the case of the complex Climate-AI nexus, to break down complexities and silos. It helps us understand what it is we are trying to solve, who we need, what capabilities they bring to the fore, and how we can maximize synergies to develop Climate AI solutions that are truly inclusive and impactful. This requires six Cs: *charting the ecosystem,* which is doing exactly what we just said above; *consulting all stakeholders*; bringing about *convergence,* that is, getting stakeholders together, making them see a common aim, and getting them to complement each other; then *collaborating* with them not just in planning but also implementation; and aiming for *cascading* impact, that is, giving it the size and scale that is needed to make an impact; and, of course, ensuring that this process is *circular* or dynamic and ongoing, especially given the constantly evolving climate and AI ecosystem.

This gives us a framework to create blueprints that can help ensure that many of the ideas contained throughout this book, and many others out there, can be planned and implemented effectively. With this framework serving as useful guidance, our vision for our solutions must also be pragmatic, that is, they should be practical, proactive, multidimensional, risk-tolerant, co-created, evidence-based and globally integrated. When systems are designed well, they also reshape incentives and behaviour, creating space for firms and institutions to converge around shared goals, collaborate meaningfully, and build credibility over time. In doing so, they turn alignment into a strategic opportunity, allowing climate- and AI-driven solutions to scale in ways that are both impactful and sustainable.

Finally, nothing would be possible without finance. The woeful state of climate finance is one of the key roadblocks to accelerated climate action globally, and ties to what we have stressed about climate and carbon equities. Climate AI will not translate from concept to real-world impact without investments at every stage, right from research to pilots, and real-world implementation and scaling up. Today, Climate AI solutions suffer from profitability gaps since they are seen to be solving a 'social' challenge; the financing landscape is fragmented, and criteria to avail finance are rigid. In an ideal world, they should be fully able to tap into all sources of finance, right from what national governments and philanthropies have to offer to grants supplied by international agencies and multilateral development banks. It is equally important for innovators, who could be startups or companies, to think about ways in which they could even make Climate AI solutions freemium—free for the public, paid for enterprise-users who benefit from those insights, such as insurance companies. To simplify access to finance, we recommend the creation of a fund of funds, government guarantees, financial rewards to Climate AI solutions that also bring about an economic impact, deployment of blended finance, democratization of global carbon markets, and channelization of global investment flows using AI, so that $1 reaches where it would have an impact worth $5, and not the other way round. Equally important, finance must increasingly be directed toward system-level enablers rather than isolated projects, supporting shared infrastructure, data platforms, and institutional capacity that allow multiple Climate AI solutions to scale together. When capital is structured to reward alignment with public priorities and long-term credibility, it not only funds innovation but also shapes behaviour across the ecosystem.

One might wonder why we decided to write a book which explores so many complexities and interdependencies, despite nudges from friends and advisors that this may end up becoming a tough read. There are a few reasons.

First, simply because no one has done this yet in such an exhaustive manner. Therefore, we decided to take it upon ourselves to do so.

Second, climate change is one of the biggest threats, and AI is one of the biggest opportunities. While climate change is a threat multiplier, AI is a force multiplier that can tackle it. Therefore, their nexus definitely warrants a thorough and comprehensive exploration.

Third, because of *vicious cycles*, which keep pushing back our progress. You might have noticed that *vicious cycles* have been a recurring theme within this book. They run through every dimension of the climate challenge, both individually and across its nexus with AI. Climate change can trigger extreme weather events such as floods or droughts, as well as conflicts. In these situations, governments often have no choice but to prioritize immediate relief and recovery, which then limits their ability to invest in long-term climate solutions. Funding for technologies that lower emissions or reinforce infrastructure becomes restricted, keeping us on a high emissions trajectory and leaving us less resilient when the next disaster occurs. This *vicious cycle* is persistent and far-reaching, and it extends deep into the world of Climate AI. For example, floods can directly disrupt the supply chains for semiconductors, which are vital for advanced technologies and data centres. Meanwhile, floods also have the potential to destroy crops, creating resource scarcity that may trigger conflicts and further destabilize essential supply chains and infrastructure, which may be thought of as an indirect disruption. By drawing attention to these interconnected links, we have tried to underline the true complexity and magnitude of the challenges ahead. Addressing them requires a multidimensional and multidisciplinary approach, calling for the entire system to come together through coordinated action.

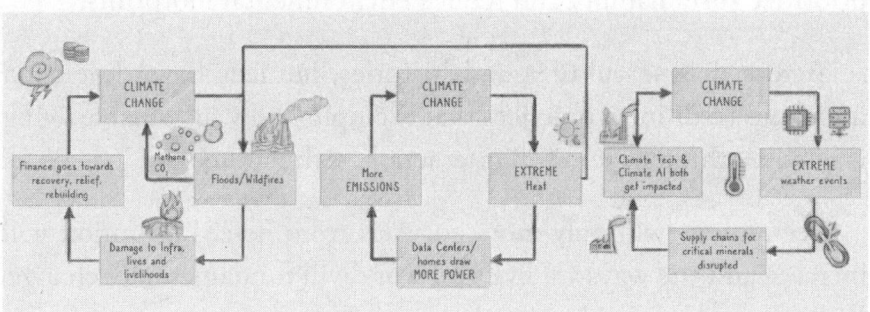

Fourth, and most importantly, you cannot solve a problem holistically unless you have explored every facet of it, every challenge, every opportunity, every interdependency, and definitely not in a single go. Unfortunately, with the pace at which climate change is progressing and the *vicious cycles* that it can create—has already started creating—we simply do not have the luxury of time for working out permutations and combinations and experimenting with solutions. There is a challenge, there is an opportunity, and there is a time-sensitive window that is fast closing.

That is precisely why our focus has been on unbundling the complexities and interdependencies and suggesting a multidimensional lens through which ecosystem-wide solutions can be developed in a collaborative manner.

In addition to this, it would be remiss of us not to acknowledge our roles as humans, as custodians of the planet, and as the future tenants of this incredible planet. While it is true that modern technologies, including AI, hold immense promise for addressing climate change, it is important to remember that these innovations ultimately complement human behaviour rather than replace or override it. Human decisions around consumption, resource use and daily habits continue to have a profound impact on the environment. Technologies can improve decision-making and efficiency, but lasting change requires fostering behaviours that prioritize sustainability and reduce environmental footprints.

Across diverse cultures and histories, humans have developed ingenious traditional practices that exemplify how sustainable living is possible through mindful interaction with nature.

Technology will only move forward from here. Disruption will increase and the ways we live and work will fundamentally change. But it is worth considering what we could continue to embrace in our daily lives, and what we can learn about harmony with nature, not only from the present, but also from those who came hundreds of years before us. The future depends not only on innovation, but on embracing behaviours and wisdom that make sustainability second-nature to our civilizations.

Throughout this book, Global South, developing countries and emerging economies are used interchangeably. While each definition has its own nuance, they broadly reflect a group of countries that share some or all common challenges and opportunities. They are still developing, so, per capita incomes continue to be low, they witness poverty; they navigate economic challenges, limited resources and social safety nets, but also high climate vulnerability. Not all are equal. Some are doing better than others, and some are doing so badly that they can be called failed nations. Yet, there are opportunities as well. The ones who are doing well are some of the fastest growing economies of the world. They have a demographic dividend on their side; they have young and aspirational populations. Perhaps, most importantly, they can technologically leapfrog—subject to the levers that we have discussed—as they are not bound by legacy constraints. It is always difficult to retrofit. It is easier to build from scratch, and these countries have opportunities to do that.

The first time many got access to banking here was straight over the internet on a smartphone. Perhaps, the new factories that are being built will be net zero from the start. But this opportunity will not exist forever and that is why this is a time-sensitive opportunity, and we should not miss this bus.

We do not claim to be experts because the world already has too many of them. What it lacks are people who take insights from everywhere and break them down into actionable insights that anyone can understand. After all, it is our shared mission, and it is one we must embrace with clarity and purpose.

One of the core goals of this book has been to strip away complexity and make the subject accessible to anyone. It does not matter if you are a student, a researcher or a policymaker. In fact, it does not even matter if climate or AI has never been your focus. If you simply have a spark of curiosity about what these ideas mean, this book is for you. That is why we chose not to drown you in jargon or technical terms. We broke everything down so clearly that some experts might curse us for underestimating their knowledge. But here is the truth. It does not matter where you come from, what you study, or what you believe in. Climate change is a global commons. The impacts ripple across countries, sectors and communities. It is a problem that belongs to all of us.

Solving this challenge demands global collaboration across sectors, disciplines and borders. Beyond policymakers, startups, businesses and international organizations who are key players, you could be a student passionate about developing Climate AI solutions. You might be a lawyer advocating for open data that is not locked behind paywalls, a historian bringing forward climate-friendly practices from societies of ages past to inspire solutions for today, an artist using your creativity to reveal the power of Climate

AI, or a diplomat working on climate negotiations. No matter your role, Climate AI holds massive potential, and every contribution matters in this fight.

Climate AI is a force multiplier for both mitigating climate change and building resilience to adapt to our warming world. By integrating diverse data, local realities and collaboration among governments, communities and innovators, AI can transform how we address and respond to climate risks. Its immense potential to protect people and the planet can break the *vicious cycle* of escalating climate damage, but realizing this requires bold vision, investment and unprecedented collaboration across all sectors. Governments must lead the way, while businesses, startups, researchers and civil society must unite to build scalable, complementary solutions for both mitigation and adaptation.

Interestingly, if we take a step back from climate and AI, many of the recommendations in this book can be applied to address challenges and leverage opportunities across multiple industries and fields. Green data centres will power AI, which is needed across sectors and industries. Open data will fuel innovation throughout diverse domains. Sovereign AI will help ensure that solutions are truly contextual and that the developing world does not end up becoming a colony of the developed world yet again. A systems lens will apply regardless of the field or challenge you are tackling. Convergence and collaboration between stakeholders will matter, and without breaking down a challenge and identifying every stakeholder, solutions will always be piecemeal. Policy support is crucial, and we must create participative policies that are practical, multidimensional, risk-tolerant and pro-innovation. Without finance, solutions even with the right policy ecosystem will never become scaled initiatives that generate real-world impact.

Today, we need multipronged solutions for multifaceted challenges. *Vicious cycles* must be transformed into virtuous ones. Threat multipliers should be countered with force multipliers. Silos must give way to collaboration, and incremental steps should be reimagined as bold leaps followed by decisive action. Climate AI falls squarely into this category, requiring all of these elements to come together. This is the only way it can become a true catalyst for collaboration, acceleration and transformation at the scale this moment demands. Climate change is the defining force of our time, and rising to its challenge requires nothing less than reimagining, acting and delivering with unprecedented urgency. We only have one shot at it, so let us get it right.

With urgency, clarity of purpose, and a spirit of collaboration, AI can turn crisis into opportunity for all humanity. The future belongs to those who embrace this moment boldly, working together to shape a sustainable, greener and more resilient world.

The future belongs to the likes of Asha, Sophie and Ori. It belongs to Mayor Ashman. It belongs to you, and it belongs to all of us.

EPILOGUE

Looking Ahead with Hope

It has been ten years since this book was published.

Mayor Ashman did not run for office a third time, and decided to retire from public life. After his retirement, he set up a foundation to support climate action and digital transformation across the Global South.

His protégé, Ori, who witnessed the devastating floods in the city, loved coding and became the first intern at the DFTCCM is now the deputy head of the department. He still loves to code, but now he is also a champion of how systems-level transformation can be ushered using an ecosystem approach. He personally led the establishment of a 2GW Green Data Centre in Climaville—one of the largest in the world. His startup has now raised funding, and is already active in four African nations.

With Mayor Ashman's vision and Ori's passion, Climaville has now become a global model of how Climate AI can be successfully implemented to help cities across the world, which now account for 80 per cent of the global population.

The conflict in Asha's village is now over, and rebuilding is underway. While her uncle, Mayor Ashman, in another part of the world, decided to retire from public life, she has just entered it. She ran in the local municipal elections and was appointed as a councillor. She campaigned on promises of renewable energy, early warning systems for disasters, technology tools to help farmers conserve water and detect crop diseases early, and AI-skilling initiatives to create livelihood opportunities for women. Interestingly, Mayor Ashman now often visits the village he migrated from decades ago and provides close guidance to Asha as she develops her ambitious plans.

Sophie, whose home was destroyed by a devastating wildfire, chose to pursue engineering and has since launched a startup that predicts and tracks wildfires. Her company makes its data and insights openly accessible, inspiring young people to use this information to create their own innovative solutions. A passionate advocate for Open Data, Sophie's recent TED talk on the subject has gone viral.

A virtual group selfie from Sophie's social media account shows her with Mayor Ashman, Ori and Asha from the Virtual Climate Action Summit on the sidelines of the G100 Summit. (Yes, the G20 has now become the G100.) While multilateral institutions across the world were working on reorienting themselves to deal with today's challenges, led by the developing world, many countries took it upon themselves to expand the G20. Already, during India's 2023 Presidency, the African Union became a member, adding fifty-five countries to the bloc. It even has a permanent secretariat now. It organized a climate action summit virtually over the metaverse so that those serious about climate action would not have to travel to the summit venue in jumbo jets that spew emissions into the atmosphere.

The world has so far managed to keep the temperature rise just below 2 degrees Celsius, but achieving that has required immense effort and much more is still needed. AI has now been officially recognized as a force multiplier in the fight against climate change, accelerating and amplifying solutions. Many years ago, the world acknowledged climate itself as a threat multiplier, something that compounded risks across sectors and regions.

For all the terrifying beginnings in this book, the ending is surprisingly hopeful. That is the beauty of this journey. Without truly understanding the scale of the problem and dissecting its complexities, pragmatic solutions would never be possible.

As for us, the world may solve one problem, but many more will crop up. We are busy with another book focused on unpacking and championing the complexities of those challenges, their nexus, and building ideas for a truly sustainable future.

NOTES

Chapter 1: An Uncertain World

1. 'Global Economy Set for Weakest Run since 2008 Outside of Recessions', Worldbank.org, 10 June 2025, https://www.worldbank.org/en/news/press-release/2025/06/10/global-economic-prospects-june-2025-press-release#:~:text=The%20turmoil%20has%20resulted%20in,global%20recession%20is%20not%20expected.
2. 'India has highest number of persons living in poverty, UN report,' Scroll.in, 20 October 2024, https://scroll.in/latest/1074675/india-has-highest-number-of-persons-living-in-poverty-un-report.
3. *Sustainable Development Goals Report 2024* (New York: United Nations, 2024), https://unstats.un.org/sdgs/report/2024/The-Sustainable-Development-Goals-Report-2024.pdf.

Chapter 2: A Warming World

1. Simon B. Vosper, 'Hazardous clear air turbulence up 55% from 1979, study shows', 8 June 2023, American Geophysical Union News, https://news.agu.org/press-release/hazardous-clear-air-turbulence-up-55-from-1979-study-shows/.

2. Mehri Khosravi et al., 'A Nation Unprepared: Extreme heat and the need for adaptation in the United Kingdom', https://uel-repository.worktribe.com/output/447835/a-nation-unprepared-extreme-heat-and-the-need-for-adaptation-in-the-united-kingdom
3. Elisa Gallo et al., 'Heat-related Mortality in Europe during 2023 and the Role of Adaptation in Protecting Health', *Nature Medicine* 2024, https://www.nature.com/articles/s41591-024-03186-1.epdf.
4. Tong Wu et al., 'Heat-related mortality in Europe during 2024 and health emergency forecasting to reduce preventable deaths', *Nature Medicine* 2024, https://www.nature.com/articles/s41591-025-03954-7.
5. Gallo et al., 'Heat-related Mortality in Europe …', 2024.
6. 'Heat-related deaths in India up by more than 50%: recent study', *The Times of India*, 2024, https://timesofindia.indiatimes.com/india/a-recent-study-has-shown-that-heat-related-deaths-in-the-country-have-gone-up-by-more-than-50-percent/articleshow/95089618.cms.
7. Ellen Gray, 'Global climate change impact on crops expected within 10 years, NASA study finds', 1 November 2021, NASA, https://climate.nasa.gov/news/3124/global-climate-change-impact-on-crops-expected-within-10-years-nasa-study-finds/.
8. *Loss and Damage in Agrifood Systems* (Rome: FAO, 2024), https://openknowledge.fao.org/handle/20.500.14283/cc8810en
9. *Landmark report at COP29: Climate-resilient transport infrastructure in South Asia 2025*, 25 March 2014, https://www.cdri.world/press-releases/cdri-and-bcg-unveil-landmark-report-at-cop29-urgent-call-for-resilient-transport-infrastructure-to-safeguard-south-asias-2-trillion-in-climate-exposed-assets
10. The European Climate Risk Assessment, published in 2024, warns that without adaptation, economic losses from coastal flooding alone could reach up to €1 trillion annually by the end of the century, https://www.eea.europa.eu/en/analysis/publications/european-climate-risk-assessment
11. '7 million premature deaths annually linked to air pollution', 25 March 2014, World Health Organization, https://www.who.int/

news/item/25-03-2014-7-million-premature-deaths-annually-linked-to-air-pollution.
12. *World Air Quality Report 2024* (Switzerland: IQAir, 2025), https://www.iqair.com/us/newsroom/waqr-2024-pr.
13. 'Report: Air Pollution Now No. 2 Killer of Children under 5', https://www.unicefusa.org/media-hub/reports/UNICEF-Air-Pollution
14. *Air Pollution Costs to Economies and Health* (Washington DC: World Bank, 2024), https://thedocs.worldbank.org/en/doc/9d5159f7d6209a13fda0768465942eab-0070012024/original/0426-Improving-Air-Quality-in-EAP-World-Bank-Daniel-Sayantan.pdf.
15. 'State of Climate in Africa highlights water stress and hazards', 08 September 2022, Wmo.int, https://wmo.int/news/media-centre/state-of-climate-africa-highlights-water-stress-and-hazards?utm_source=chatgpt.com.
16. 'India's water availability projections', 2024, https://www.pib.gov.in/PressReleaseIframePage.aspx?PRID=2002726&utm_.
17. Ajit Niranjan, 'Climate change doubled chance of extreme rain in Europe in September', *The Guardian*, 25 September 2024, https://www.theguardian.com/environment/2024/sep/25/global-heating-doubled-chance-of-extreme-rain-in-europe-in-september.
18. Watts, Nick et al. "The 2024 Lancet Countdown on health and climate change: early signs." https://www.thelancet.com/journals/lancet/article/PIIS0140-6736(24)01822-1/abstract
19. 'In 2016, A Freak Anthrax Outbreak Swept Across Siberia—And Thawing Permafrost Means It Could Happen Again', https://www.forbes.com/sites/scotttravers/2025/03/16/in-2016-a-freak-anthrax-outbreak-swept-across-siberia-and-thawing-permafrost-means-it-could-happen-again/
20. Damian Carrington, 'Triple whammy of hottest ever years risks "irreversible damage", says UN', 6 November 2025, *The Guardian*, https://www.theguardian.com/environment/2025/nov/06/triple-whammy-of-hottest-ever-years-risks-irreversible-damage-says-un.
21. 'Greenhouse gas concentrations hit record high. Again.', 2025, WMO, https://wmo.int/news/media-centre/greenhouse-gas-concentrations-hit-record-high-again.

22. 'Arctic sea ice decline', 2024, National Snow and Ice Data Centre, https://nsidc.org/learn/parts-cryosphere/sea-ice.
23. Fisher, Max. 'The planet has lost half of coral reefs since 1950', *Smithsonian Magazine*, 2024, https://www.smithsonianmag.com/science-nature/the-planet-has-lost-half-of-coral-reefs-since-1950-180978701/.
24. 'Global mean sea level rise', 2024, NASA, https://sealevel.nasa.gov/understanding-sea-level/key-indicators/global-mean-sea-level.

Chapter 3: Tech on Thin Ice

1. 'Influencer Market Size, Share, Growth, and Industry Analysis ...', 10 November 2025, Business Research Insights, https://www.businessresearchinsights.com/market-reports/influencer-market-10760.1
2. 'What is Microsoft's "blue screen of death?" Here's what it means and how to fix it', https://www.cbsnews.com/news/microsoft-crowdstrike-outage-blue-screen-of-death-how-to-fix/
3. Josh Lepawsky and Helen Millman, 'How climate change and water stress is risking the semiconductor supply chain', 2 December 2024, Weforum.org, https://www.weforum.org/stories/2024/12/how-climate-change-and-water-stress-is-risking-the-semiconductor-supply-chain/.
4. Karen Chiu, Hannah Wang and Sam Phillips, 'A third of global chip supply threatened by climate change and drought by 2035', 8 July 2025, *South China Morning Post*, https://www.scmp.com/tech/tech-trends/article/3317237/third-global-chip-supply-threatened-climate-change-and-drought-2035-pwc.
5. Alyssa Bersine, 'Reducing data center peak cooling demand and energy costs with underground thermal energy storage', 17 January 2025, NERL, https://www.nrel.gov/news/detail/program/2025/reducing-data-center-peak-cooling-demand-and-energy-costs-with-underground-thermal-energy-storage.
6. *Securing Clean Energy Technology Supply Chains* (Australia: International Energy Agency, 2022), https://iea.blob.core.

windows.net/assets/0fe16228-521a-43d9-8da6-bbf08cc9f2b4/SecuringCleanEnergyTechnologySupplyChains.pdf?

Chapter 4: The Geopolitics of Climate Change

1. Emilie Yam, 'Does climate change cause conflict?', 2 June 2021, International Growth Centre, https://www.theigc.org/blogs/climate-priorities-developing-countries/does-climate-change-cause-conflict.
2. 'How climate-induced conflict is shaping rural Nigeria', https://www.cgiar.org/news-events/news/how-climate-induced-conflict-is-shaping-rural-nigeria
3. 'US denies Mexico's request for special delivery channel for Colorado River water', https://www.reuters.com/world/us/us-denies-mexicos-request-special-delivery-channel-colorado-river-water-2025-03-20/
4. 'Landmark report on impacts of disappearing snow and ice in the Hindu Kush Himalaya – current emissions path threatens two billion people and is accelerating species extinction', 20 June 2023, International Centre for Integrated Mountain Development, https://www.icimod.org/press-release/landmark-report-on-impacts-of-disappearing-snow-and-ice-in-the-hindu-kush-himalaya-current-emissions-path-threatens-two-billion-people-and-is-accelerating-species-extinction/.
5. Devon Ryan, 'Stanford-led study investigates how much climate change affects the risk of armed conflict', 12 June 2019, Standford Report, https://news.stanford.edu/stories/2019/06/climate-change-cause-armed-conflict.
6. 'Mineral production to soar as demand for clean energy increases', 11 May 2020, World Bank, https://www.worldbank.org/en/news/press-release/2020/05/11/mineral-production-to-soar-as-demand-for-clean-energy-increases
7. 'Executive Summary', Global Critical Minerals Outlook 2025, https://www.iea.org/reports/global-critical-minerals-outlook-2025/executive-summary
8. 'Summary for Policymakers', IPCC's Sixth Assessment, https://www.ipcc.ch/report/ar6/wg2/chapter/summary-for-policymakers/.

9. Drought, World Health Organization, https://www.who.int/health-topics/drought#tab=tab_1.
10. 'Climate Security Mechanism', 2024, United Nations, https://www.un.org/climatesecuritymechanism/en/about/climate-security-mechanism
11. 'World Bank Group Announces Sangbu Kim as Vice President for Digital Transformation', https://www.worldbank.org/en/news/press-release/2024/07/30/world-bank-group-announces-sangbu-kim-as-vice-president-for-digital-transformation
12. 'Voice of the Global South Summit', Embassy of India, Seoul, http://www.indembassyseoul.gov.in/voice-global-south-summit

Chapter 5: Rethinking Climate Equity

1. 'India should be among 1st nations to industrialise without carbonizing the world: Amitabh Kant', https://economictimes.indiatimes.com/news/india/india-should-be-among-1st-nations-to-industrialise-without-carbonizing-the-world-amitabh-kant/articleshow/100306509.cms
2. 'Developed Nations Consuming Over 80% Of Carbon Budget, Centre Tells Rajya Sabha', https://www.ndtv.com/india-news/developed-nations-consuming-over-80-of-carbon-budget-centre-tells-rajya-sabha-4287870
3. 'CO_2 emissions per capita', https://ourworldindata.org/grapher/co-emissions-per-capita
4. 'Analysis: Which countries are historically responsible for climate change?', https://www.carbonbrief.org/analysis-which-countries-are-historically-responsible-for-climate-change/
5. 'India's historical cumulative emissions and per capita emissions are very low despite being home to more than 17% of the global population', https://www.pib.gov.in/PressReleasePage.aspx?PRID=1842619&utm&reg=3&lang=2
6. 'Weather-related disasters increase over past 50 years, causing more damage but fewer deaths', 30 August 2021, World Meteorological Organization, https://wmo.int/media/news/weather-related-

disasters-increase-over-past-50-years-causing-more-damage-fewer-deaths
7. 'Climate Finance in the negotiations', https://unfccc.int/topics/climate-finance/the-big-picture/climate-finance-in-the-negotiations?
8. 'COP27 Reaches Breakthrough Agreement on New "Loss and Damage" Fund for Vulnerable Countries', https://unfccc.int/news/cop27-reaches-breakthrough-agreement-on-new-loss-and-damage-fund-for-vulnerable-countries
9. 'Fund for responding to Loss and Damage', https://unfccc.int/fund-for-responding-to-loss-and-damage; https://unfccc.int/sites/default/files/resource/B.3%20Compendium_of_decisions.pdf
10. 'Belém COP30 delivers climate finance boost and a pledge to plan fossil fuel transition', https://news.un.org/en/story/2025/11/1166433
11. 'A new climate finance goal is on the horizon', https://unctad.org/news/new-climate-finance-goal-horizon-how-can-developing-countries-benefit#:~:text=Based%20on%20modelled%20projections%20using%20the%20United%20Nations,from%202025%20and%20some%20%241.8%20trillion%20by%202030.
12. 'COP29 UN Climate Conference Agrees to Triple Finance to Developing Countries, Protecting Lives and Livelihoods', https://unfccc.int/news/cop29-un-climate-conference-agrees-to-triple-finance-to-developing-countries-protecting-lives-and
13. 'Finance and investment for climate goals - OECD', https://www.oecd.org/en/topics/policy-issues/finance-and-investment-for-climate-goals

Chapter 6: Challenges as Opportunities
1. 'WSIS+20 HIGH-LEVEL EVENT 2025', https://women20.org/wp-content/uploads/2025/12/g20-Indonesia-declaration-data.pdf
2. 'G20 New Delhi Leaders' Declaration', https://www.g20.in/content/dam/gtwenty/gtwenty_new/document/G20-New-Delhi-Leaders-Declaration.pdf
3. 'G20 Rio de Janeiro Leaders' Declaration - G20 Research Group', https://www.g20.utoronto.ca/2024/241118-declaration.html

4. 'G20 South Africa Summit: Leaders' Declaration', https://www.consilium.europa.eu/media/yfyd2czp/declaration-g20-south-africa-summit-22-23-november.pdf
5. 'G20 South Africa Summit: Leaders' Declaration', https://www.pib.gov.in/PressReleasePage.aspx?PRID=2129952&reg=3&lang=2
6. 'All Press Release: Press Information Bureau', https://www.pib.gov.in/PressReleasePage.aspx?PRID=2144627&reg=3&lang=2
7. 'Brazil Reaches 85% Clean Power Targets 90% by 2030', https://www.riotimesonline.com/brazil-reaches-85-clean-power-targets-90-by-2030-amid-biofuel-push/#:~:text=Brazil%E2%80%99s%20Ministry%20of%20Mines%20and%20Energy%20confirmed%20the,new%20solar%20and%20wind%20projects%20added%20in%202024.
8. Malgorzata Wiatros-Motyka et al., 'Global Electricity Mid-Year Insights 2025: Renewables overtake coal and gas for the first time', 7 October 2025, Ember Energy, https://ember-energy.org/latest-insights/global-electricity-mid-year-insights-2025/.

Chapter 7: On the Cutting Edge

1. Xiaoying You, 'The "typhoon-proof" wind farms powering China's coast', 7 October 2025, BBC Future, https://www.bbc.com/future/article/20251006-the-typhoon-proof-wind-farms-powering-chinas-coast
2. 'A.P.'s largest integrated renewable energy project likely to become operational by June', https://www.thehindu.com/news/national/andhra-pradesh/aps-largest-integrated-renewable-energy-project-likely-to-become-operational-by-june/article67765691.ece#:~:text=Greenko%20is%20the%20developer%20of%20Pinnapuram%20IREP%2C%20which,components%20that%20can%20supply%20schedulable%20power%20on%20demand.
3. 'Energy Vault's First Grid-Scale Gravity Energy Storage System Is Near Complete', https://singularityhub.com/2023/08/09/energy-vaults-first-grid-scale-gravity-energy-storage-system-is-near-complete/

Chapter 8: The AI Revolution

1. 'Demis Hassabis & John Jumper awarded Nobel Prize in Chemistry', https://deepmind.google/blog/demis-hassabis-john-jumper-awarded-nobel-prize-in-chemistry/#:~:text=This%20morning%2C%20Co-founder%20and%20CEO%20of%20Google%20DeepMind,structure%20of%20proteins%20from%20their%20amino%20acid%20sequences.
2. 'The state of AI: How organizations are rewiring to capture value', 12 March 2025, McKinsey & Company, https://www.mckinsey.com/capabilities/quantumblack/our-insights/the-state-of-ai-how-organizations-are-rewiring-to-capture-value
3. 'The state of AI in early 2024: Gen AI adoption spikes and starts to generate value', 30 May 2024, McKinsey & Company, https://www.mckinsey.com/capabilities/quantumblack/our-insights/the-state-of-ai-2024
4. 'Sizing the prize: What's the real value of AI for your business and how can you capitalise?' 2017, PwC, https://www.pwc.com/gx/en/issues/analytics/assets/pwc-ai-analysis-sizing-the-prize-report.pdf
5. 'The economic potential of generative AI: The next productivity frontier', 14 June 2023, McKinsey & Company, https://www.mckinsey.com/capabilities/tech-and-ai/our-insights/the-economic-potential-of-generative-ai-the-next-productivity-frontier
6. 'Google to invest $15 billion in AI data centre in biggest India investment', https://economictimes.indiatimes.com/tech/technology/google-to-invest-10-billion-in-data-centre-in-south-india/articleshow/124542008.cms
7. 'Microsoft invests US$17.5 billion in India to drive AI diffusion at population scale', https://news.microsoft.com/source/asia/2025/12/09/microsoft-invests-us17-5-billion-in-india-to-drive-ai-diffusion-at-population-scale/
8. 'L&T rebrands data centre business as Larsen & Toubro-Vyoma', 26 November 2025, Economic Times, https://www.economictimes.com/news/company/corporate-trends/lt-rebrands-data-centre-business-as-larsen-toubro-vyoma-expands-digital-push/articleshow/125583773.cms

9. 'Google's 'TPU' chip puts OpenAI on alert and shakes Nvidia investors', https://www.ft.com/content/d8585870-17a5-43a0-95ef-cbebb1995107
10. 'AI and global labor skill evolution', 2025, LinkedIn Economic Graph, https://economicgraph.linkedin.com
11. 'How did we get a dozen '1000-year floods' in 3 days?', https://www.accuweather.com/en/weather-news/how-did-we-get-a-dozen-1000-year-floods-in-3-days/1792443
12. 'European heatwave caused 2,300 deaths, scientists estimate', https://www.reuters.com/sustainability/cop/european-heatwave-caused-2300-deaths-scientists-estimate-2025-07-09/
13. 'Extreme weather impacts cascading "from the Andes to the Amazon"', https://news.un.org/en/story/2025/03/1161626

Chapter 9: Predicting the Unpredictable

1. 'A big step for flood forecasts in India and Bangladesh', https://blog.google/technology/ai/flood-forecasts-india-bangladesh/
2. Yimou Lee, 'Storm Bebinca Approaches Taiwan, Uses AI to Predict Typhoon Paths', 13 September 2024, Reuters, https://www.reuters.com/technology/artificial-intelligence/storm-bebinca-approaches-taiwan-uses-ai-predict-typhoon-paths-2024-09-13/.
3. 'Lisbon's City-Scale Digital Twins for Flood Resilience', Geospatial World, https://geospatialworld.net/prime/case-study/aec/lisbons-city-scale-digital-twins-for-flood-resilience-2/
4. Lindsey Jean Schueman, 'How Ryan Honary developed AI-powered wildfire sensors to protect communities', 13 March 2025, One Earth, https://www.oneearth.org/climate-hero-ryan-honary/.
5. Nina Raffio, 'USC scientists use AI to predict a wildfire's next move', 22 July 2024, USC Today, https://today.usc.edu/using-ai-to-predict-wildfires/.
6. Chakib Jenane, 'Is Artificial Intelligence the future of farming? Exploring opportunities and challenges in Sub-Saharan Africa', 12 March 2025, World Bank Blogs, https://blogs.worldbank.org/en/agfood/artificial-interlligence-in-the-future-of-sub-saharan-africa-far.

7. Giselle Ombay, 'UP Flood Map Shows AI Importance in Disaster Preparedness, Says UP President', 24 October 2024, GMA Network, https://www.gmanetwork.com/news/scitech/technology/924802/up-flood-map-shows-ai-importance-in-disaster-preparedness-says-up-president/story/.
8. Samuel Chege Maina, 'Extending ClimaX for Flood Forecasting', 9 April 2025, Youtube video, Microsoft Research, https://www.microsoft.com/en-us/research/video/extending-climax-for-flood-forecasting/?msockid=27679553a6586de504fd80e1a7aa6cdb
9. 'One Concern Launches First-Ever Digital Twin to Build Climate Resilience', 22 February 2022, *Business Wire*, https://www.businesswire.com/news/home/20220222005217/en/One-Concern-Launches-First-Ever-Digital-Twin-to-Build-Climate-Resilience
10. Harry Booth, 'Thomas Njeru', 2 October 2024, *Time*, https://time.com/7023557/thomas-njeru/.
11. '7 million premature deaths annually linked to air pollution', https://www.who.int/news/item/25-03-2014-7-million-premature-deaths-annually-linked-to-air-pollution
12. Yael Maguire and Miriam Daniel, 'Using Google's AI and Local Ecosystem to Generate Actionable Air Quality Insights in India with Air View', 20 November 2024, Google Blog, https://blog.google/intl/en-in/company-news/using-googles-ai-and-local-ecosystem-to-generate-actionable-air-quality-insights-in-india-with-air-view/
13. 'Air pollution in South Africa: Affordable new devices use AI to monitor hotspots in real time', 16 August 2024, Green Building Africa, https://www.greenbuildingafrica.co.za/air-pollution-in-south-africa-affordable-new-devices-use-ai-to-monitor-hotspots-in-real-time/
14. 'Bring time back to care with AI-assisted applications for dengue in Vietnam', 16 June 2025, OUCRU, https://www.oucru.org/bring-time-back-to-care/.
15. 'Transforming Malaria Control in Benin with AI-Powered Diagnostic Solutions', 22 May 2025, NOUL, https://noul.com/en/board_news_blog/transforming-malaria-control-in-benin/.

16. Nell Lewis, 'How drones and AI are protecting the Brazilian rainforest', 12 November 2025, CNN, https://edition.cnn.com/world/americas/regreen-earthshot-tech-rainforest-brazil-spc.
17. Samira Njoya, 'Kenya: AstraZeneca unveils AI-monitored initiative to plant 6mln trees', 5 December 2023, WeAreTech, https://www.wearetech.africa/en/fils-uk/news/tech/kenya-astrazeneca-unveils-ai-monitored-initiative-to-plant-6mln-trees.

Chapter 10: Decarbonizing the Planet

1. Loadshedding is the deliberate, controlled shutdown of electricity to parts of a power grid to prevent a total system collapse when demand exceeds supply. It's a protective measure, often implemented as 'rolling blackouts', to balance the load by intentionally reducing demand when a power grid is under extreme stress from high electricity consumption or unexpected power failures.
2. Heidi Opdyke, 'Surtrac Allows Traffic To Move at the Speed of Technology', 25 October 2019, Carnegie Mellon University News, https://www.cmu.edu/news/stories/archives/2019/october/traffic-moves-at-speed-of-technology.html.
3. Matt Villano, 'How Uber Freight is leveraging AI to make truck routes more efficient', 10 April 2025, *Business Insider*, https://www.businessinsider.com/ai-trucking-logistics-uber-freight-tech-optimize-routes-2025-4.
4. Project Sunroof-Google, https://sunroof.withgoogle.com/
5. Joseph E. Harmon, 'AI helps whittle down candidates for hydrogen carriers in liquid form from billions to about 40', 10 January 2024, Argonne National Laboratory, https://www.anl.gov/article/ai-helps-whittle-down-candidates-for-hydrogen-carriers-in-liquid-form-from-billions-to-about-40.
6. 'Microsoft and PNNL Use AI to Discover New Compound for Carbon Capture', 25 May 2025, NBC Right Now, https://www.nbcrightnow.com/news/microsoft-and-pnnl-use-ai-to-discover-new-compound-for-carbon-capture/article_9ba90dd6-29bd-4e3b-a6ed-b5ed350bd0b6.html.

7. Todd Bush, 'First-of-its-kind AI-powered tech captures carbon using heat from data centers', 14 April 2025, Decarbon Fuse, https://decarbonfuse.com/posts/first-of-its-kind-ai-powered-tech-captures-carbon-using-heat-from-data-centers.
8. 'Amrize and Meta Partner on First-of-its-kind AI-Optimized Advanced Concrete Mix for Data Center in Minnesota', Amrize, https://www.amrize.com/us/en/newsroom/amrize-and-meta-partner-on-ai-optimized-concrete.html
9. Richard Evans and Jim Gao, 'DeepMind AI reduces Google data centre cooling bill by 40%', 20 July 2016, DeepMind Google, https://deepmind.google/discover/blog/deepmind-ai-reduces-google-data-centre-cooling-bill-by-40/.
10. 'Hard-to-Abate Sectors and Emissions-CitiGPS', https://www.citigroup.com/global/insights/hard-to-abate-sectors-and-emissions
11. 'AI optimises scrap flow at Hamburg mill', 7 December 2022, EuroMetal, https://eurometal.net/ai-optimises-scrap-flow-at-hamburg-mill/
12. 'Heidelberg Materials improves performance by integrating Carbon Re AI', 7 October 2024, Carbon Re, https://carbonre.com/heidelberg-materials-improves-performance-by-integrating-carbon-re-ai
13. Sharm-el-Sheikh, 'World's largest carbon program pilots digital measuring of forest carbon', 16 November 2022, Pachama, https://pachama.com/blog/worlds-largest-carbon-program-pilots-digital-measuring-of-forest-carbon/
14. 'Climate TRACE - Emissions Mapping', Climate TRACE, https://climatetrace.org

Chapter 11: Intelligence at Scale

1. EarthScan Platform, Cervest, https://www.earth-scan.com/
2. 'AI and Blockchain-driven Capital Markets Connectivity to Green Real World Assets', ZERO13, https://zero13.net/bridging-the-climate-financing-gap-with-ai-and-blockchain-driven-capital-markets-connectivity-to-green-real-world-assets/

Chapter 12: From Dilemma to Design

1. IMF Blog, https://www.imf.org/en/blogs/chart-of-the-week
2. IMF Blog, https://www.imf.org/en/blogs/chart-of-the-week

Chapter 13: Data Centres of the Future

1. 'Hyperscalers to Account for Half the $1.2 Trillion Global Data Center Capex by 2029', 6 August 2025, DellOro Group, https://www.delloro.com/news/data-center-capex-to-grow-at-21-percent-cagr-through-2029/
2. 'AI is poised to drive 160% increase in data center power demand', https://www.goldmansachs.com/insights/articles/AI-poised-to-drive-160-increase-in-power-demand
3. 'Investment in Raxio Group for African Data Centers', World Bank, https://www.worldbank.org/en/news/feature/2022/03/08/africa-the-digital-frontier
4. 'Investment to Build AI Hub in India', Google Blog, https://www.blog.google/inside-google/infrastructure/expanding-data-centers-across-india/
5. D. Mytton, 'Data centre water consumption', *Nature Partner Journals Clean Water*, Volume 4, No. 11 (2021), https://www.nature.com/articles/s41545-021-00101-w#citeas
6. 'World Bank backs Africa digital data push with $100 million Raxio deal', https://www.reuters.com/world/africa/world-bank-backs-africa-digital-data-push-with-100-million-raxio-deal-2025-04-03/
7. 'AI is set to drive surging electricity demand from data centres while offering the potential to transform how the energy sector works', https://www.iea.org/news/ai-is-set-to-drive-surging-electricity-demand-from-data-centres-while-offering-the-potential-to-transform-how-the-energy-sector-works
8. 'Uptime Institute Global Data Center Survey 2024', https://datacenter.uptimeinstitute.com/rs/711-RIA-145/images/2024.GlobalDataCenterSurvey.Report.pdf
9. Meta. 'Data Center in Luleå, Sweden - Free Air Cooling'. https://about.fb.com/news/2021/02/lulea-data-center/

10. 'Luleå goes live - Meta Data Centers', https://datacenters.atmeta.com/2013/06/lulea-goes-live/
11. 'Google Expands Data Centre Footprint with Finnish Land Deal', https://datacentremagazine.com/articles/google-expands-data-centre-footprint-with-finnish-land-deal
12. 'Starcloud runs AI model in space', 12 December 2025, Data Center Dynamics, https://www.datacenterdynamics.com/en/news/starcloud-runs-ai-model-in-space/
13. 'Data centres in space? Jeff Bezos thinks it's possible', 4 October 2025, Reuters, https://www.reuters.com/business/energy/data-centres-space-jeff-bezos-thinks-its-possible-2025-10-03/
14. 'Microsoft's undersea datacenter helps the hunt for a COVID-19 vaccine, 16 June 2020', Microsoft, https://news.microsoft.com/innovation-stories/project-natick/
15. 'Project Natick Phase 2 - Microsoft', https://natick.research.microsoft.com/
16. 'AWS expands in Latin America with first Chile cloud region', https://dig.watch/updates/aws-expands-in-latin-america-with-first-chile-cloud-region
17. 'Persistent drought is drying out Chile's drinking water', https://www.reuters.com/world/americas/persistent-drought-is-drying-out-chiles-drinking-water-2024-03-20/
18. 'Africa Data Centres, DPA break ground on 12 MW solar farm in South Africa', https://www.reuters.com/business/energy/africa-data-centres-dpa-break-ground-12-mw-solar-farm-south-africa-2024-04-08/
19. 'Fully Green Data Centre Unveiled in Kenya', https://cioafrica.co/africa-first-green-data-centre-launched-in-kenya/
20. 'MainOne Enters Cote d'Ivoire Expanding Its Data Center Footprint', https://www.datacenterknowledge.com/data-center-construction/mainone-enters-cote-d-ivoire-expanding-its-data-center-footprint; 'Energy efficiency is key to the sustainability of data centres in Africa', https://mainone.africa-newsroom.com/press/energy-efficiency-is-key-to-the-sustainability-of-data-centres-in-africa?lang=en

21. 'Energy efficiency is key to the sustainability of data centres in Africa', https://www.ytl.com/press-releases/ytl-green-data-center-park-launches-in-johor-the-first-integrated-data-center-park-powered-by-renewable-solar-energy-in-malaysia-2/
22. 'AirTrunk Powers Up Sustainability in Malaysia with Solar Initiative', https://w.media/airtrunk-powers-up-sustainability-in-malaysia-with-solar-initiative
23. 'Maharashtra to develop 1.5 GW Green DC Parks', https://w.media/maharashtra-to-develop-1-5-gw-green-dc-parks/
24. 'Ovhcloud Enterprise Hosting', https://corporate.ovhcloud.com/asia/newsroom/new-datacentre-sydney/
25. 'DeepMind AI Reduces Google Data Centre Cooling Bill by 40%', https://deepmind.google/blog/deepmind-ai-reduces-google-data-centre-cooling-bill-by-40/
26. 'AWS Reaches 53% of Water Positive Goal', https://mexicobusiness.news/ecommerce/news/aws-reaches-53-water-positive-goal
27. 'Our first offsite heat recovery project lands in Finland', https://blog.google/around-the-globe/google-europe/our-first-offsite-heat-recovery-project-lands-in-finland/
28. 'Brookfield and Microsoft Collaborating to Deliver Over 10.5 GW of New Renewable Power Capacity Globally', https://bep.brookfield.com/press-releases/bep/brookfield-and-microsoft-collaborating-deliver-over-105-gw-new-renewable-power
29. 'Google, NV Energy Plan To Use Geothermal Power For Data Centers', https://www.saurenergy.com/solar-energy-news/google-nv-energy-plan-to-use-geothermal-power-for-data-centers
30. 'Alphabet Announces Agreement to Acquire Intersect Power', 22December 2025, Alphabet Investor Relations, https://abc.xyz/investor/news/news-details/2025/Alphabet-Announces-Agreement-to-Acquire-Intersect-Power/

Chapter 14: DPI and Data Commons for Climate AI

1. 'Sweden is on its way to becoming a cashless society', https://www.weforum.org/stories/2017/09/sweden-becoming-cashless-society

2. 'Digital Public Infrastructure: A Foundation for DigitalTransformation', United Nations Development Programme (UNDP), https://www.undp.org/publications/digital-public-infrastructure
3. *India Stack and UPI: Driving Financial Inclusion Through DPI*, Government of India, https://www.npci.org.in/product-overview/upi-product-overview
4. 'UPI In October: PhonePe, Google Pay's UPI Lead Softens Amid Record High Volumes', https://inc42.com/buzz/upi-in-october-phonepe-google-pays-upi-lead-softens-amid-record-high-volumes/
5. 'The Transformative Power of Ethiopia's Digital ID: Unlocking a better future for all', https://www.worldbank.org/en/news/feature/2025/02/27/the-transformative-power-of-ethiopia-afe-digital-id-unlocking-a-better-future-for-all
6. 'Approaches to Digital Public Infrastructure in the Global South', https://www.csis.org/analysis/approaches-digital-public-infrastructure-global-south
7. State Enterprise, https://se.diia.gov.ua/en/trembita
8. Digital Government Payment Platforms for Social Protection, Zambia Government, https://www.zambiagovernment.gov.zm/digital-transformation

Chapter 15: Decolonizing AI

1. 'Chinese AI start-up DeepSeek pushes US rivals with R1 model upgrade', https://www.reuters.com/world/china/chinas-deepseek-releases-an-update-its-r1-reasoning-model-2025-05-29/
2. 'Explainer: What is DeepSeek and why is it disrupting the AI sector?', https://www.reuters.com/technology/artificial-intelligence/what-is-deepseek-why-is-it-disrupting-ai-sector-2025-01-27
3. 'The Value of Open Source Software', https://www.hbs.edu/ris/Publication%20Files/24-038_51f8444f-502c-4139-8bf2-56eb4b65c58a.pdf
4. Patricia DeLacey, 'Biases in Large Image-Text AI Models FavorWealthier Western Perspectives' 8 December 2023, University

of Michigan, https://news.umich.edu/biases-in-large-image-text-ai-model-favor-wealthier-western-perspectives/
5. Jake Okechukwu Effoduh, 'A Global South Perspective on Explainable AI', 30 April 2024, Carnegie Endowment for International Peace, https://carnegieendowment.org/research/2024/04/a-global-south-perspective-on-explainable-ai?lang=en
6. 'Examination of Cultural Bias in Large Language Models', KTH Royal Institute of Technology and University of Pennsylvania, https://watermark02.silverchair.com/pgae346.pdf
7. 'Addressing AI Bias in Maternal Healthcare in Southern Africa', https://www.mozillafoundation.org/en/blog/addressing-ai-bias-in-maternal-healthcare-in-southern-africa
8. Mozilla Foundation (2023). Addressing AI Bias in Maternal Healthcare in Southern Africa. Mozilla Foundation. Available at: https://www.mozillafoundation.org/de/blog/addressing-ai-bias-in-maternal-healthcare-in-southern-africa/
9. Nidhi Jamwal, 'Data Drought: The Challenge of AI Weather Forecasting in India's Himalayan Region', 26 April 2024, Dialogue Earth, https://dialogue.earth/en/climate/data-drought-the-challenge-of-ai-weather-forecasting-in-india/
10. 'Red Hat CEO Helps Turn Open Source Software Into $2 Billion Business', https://www.forbes.com/sites/brucerogers/2015/06/18/red-hat-ceo-helps-turn-open-source-software-into-2-billion-business/
11. 'IBM and NASA are building an AI foundation model for weather and climate', https://research.ibm.com/blog/foundation-model-weather-climate
12. 'The Value of Open Source Software - Working Paper', https://www.hbs.edu/faculty/Pages/item.aspx?num=65230

ABOUT THE AUTHORS

Amitabh Kant is a governance reformer and public-policy change agent who has led major national initiatives and transformative reforms over a distinguished forty-five-year career at the highest levels of the Government of India. He is widely regarded as a key architect of India's modern economic and brand identity.

He is the chancellor of NIIT University and currently serves as a senior adviser to Sumitomo Mitsui Banking Corporation (SMBC), Fairfax and Warburg Pincus. He is on the boards of Larsen and Toubro (L&T), HCL Tech, ITC and IndiGo and is a trustee at the Council on Energy, Environment and Water (CEEW), a member of the advisory board at The Convergence Foundation, and a member of the board of governors at the Indian Council for Research on International Economic Relations (ICRIER).

He previously served as the Prime Minister's G20 sherpa during the Indian G20 Presidency in 2023. As sherpa, he delivered the landmark New Delhi Leader's Declaration, which carried key ambitions and commitments around AI and Climate, the latter through the Green Development Pact.

Prior to G20, he served as CEO of the National Institution for Transforming India (NITI Aayog), which is the Prime Minister-led apex policy-planning agency of the Government of India. As CEO, Kant led several initiatives around Climate and AI including India's first National AI Strategy, the setting up of the National EV Mission, the National Green Hydrogen Mission, the National Mission for Battery Storage and the National Circular Economy Mission. He also drove some of India's key policy programmes such as the Aspirational Districts Programme, the Production Linked Incentive (PLI) schemes to boost manufacturing, and the Asset Monetisation programme.

He has also held numerous key positions with the Government of India, including as secretary of the Department of Industrial Policy and Promotion, during which he launched the Make in India and Startup India initiatives. He also served as the joint secretary of Tourism, leading the Incredible India campaign.

He is the author of several books including *Made in India: 75 Years of Business and Enterprise*; *Incredible India 2.0: Synergies for Growth and Governance*; *Branding India: An Incredible Story*; and *How India Scaled Mt. G20: The Inside Story of the G20 Presidency*. He is the co-author of *Elephant Moves: India's New Place in the World* and the editor of *The Path Ahead: Transformative Ideas for India*.

Kant graduated with a degree in Economics from St. Stephen's College, Delhi, and earned an MA in International Relations from Jawaharlal Nehru University. He was a Chevening Scholar and is the recipient of The Order of the Rising Sun, Gold and Silver Star. He was honoured with the Sir Edmund Hillary Fellowship.

Siddharth Sinha specializes in the interplay of AI, climate, energy and digital public infrastructure (DPI). He previously worked at Google, where he set up a cross-company initiative aimed at accelerating AI × Climate solutions in the Global South. He also co-conceptualized an air quality management solution operational in 150+ cities across India and Brazil.

Prior to that, he worked at the G20 Secretariat where he served as the adviser to India's G20 sherpa during the Indian presidency, providing strategic support towards the emergence of the Green Development Pact and African Union's inclusion into the G20. Before G20, Siddharth worked with the National Institution for Transforming India (NITI Aayog) as the policy specialist for Climate, Transport, Digital and Energy, and the chief of staff to the CEO. During this time, he contributed to national climate policy matters including financing, transition pathways, clean mobility, circular economy, cleantech and DPI. Siddharth nationally coordinated two of India's key decarbonization initiatives which, among others, resulted in the development of India's first life cycle assessment model for transport emissions. He contributed to India's COP26 strategy, the setting up of the country's National EV Mission, and managed an initiative that led to the emergence of national-scale digital inclusion platforms, built in partnership with the private sector. Siddharth also served as India's representative to OECD's International Transport Forum.

Siddharth holds a B.Tech. degree in Electrical Engineering from Vellore Institute of Technology (VIT), India, and a Master of Public Administration in International Development (MPA-ID) from the London School of Economics and Political Science, UK. He is a 2024 Climate Fellow at Yale University, and was a Chevening Science, Research and Innovation Fellow at the University of Oxford. He is currently an associate vice president at Greenko, a leading renewables and green molecules company.

Siddharth is a mountaineer and a TEDx speaker.

HarperCollins *Publishers* India

At HarperCollins India, we believe in telling the best stories and finding the widest readership for our books in every format possible. We started publishing in 1992; a great deal has changed since then, but what has remained constant is the passion with which our authors write their books, the love with which readers receive them, and the sheer joy and excitement that we as publishers feel in being a part of the publishing process.

Over the years, we've had the pleasure of publishing some of the finest writing from the subcontinent and around the world, including several award-winning titles and some of the biggest bestsellers in India's publishing history. But nothing has meant more to us than the fact that millions of people have read the books we published, and that somewhere, a book of ours might have made a difference.

As we look to the future, we go back to that one word—a word which has been a driving force for us all these years.

Read.

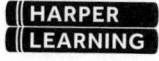